SpringerBriefs in Research and Innovation Governance

Editors-in-Chief

Doris Schroeder, Centre for Professional Ethics, University of Central Lancashire, Preston, Lancashire, UK

Konstantinos Iatridis, School of Management, University of Bath, Bath, UK

SpringerBriefs in Research and Innovation Governance present concise summaries of cutting-edge research and practical applications across a wide spectrum of governance activities that are shaped and informed by, and in turn impact research and innovation, with fast turnaround time to publication. Featuring compact volumes of 50 to 125 pages, the series covers a range of content from professional to academic. Monographs of new material are considered for the SpringerBriefs in Research and Innovation Governance series. Typical topics might include: a timely report of state-of-the-art analytical techniques, a bridge between new research results, as published in journal articles and a contextual literature review, a snapshot of a hot or emerging topic, an in-depth case study or technical example, a presentation of core concepts that students and practitioners must understand in order to make independent contributions, best practices or protocols to be followed, a series of short case studies/debates highlighting a specific angle. SpringerBriefs in Research and Innovation Governance allow authors to present their ideas and readers to absorb them with minimal time investment. Both solicited and unsolicited manuscripts are considered for publication.

Bernd Carsten Stahl · Doris Schroeder ·
Rowena Rodrigues

Ethics of Artificial Intelligence

Case Studies and Options for Addressing Ethical Challenges

 Springer

Bernd Carsten Stahl
School of Computer Science
University of Nottingham
Nottingham, UK

Centre for Computing and Social
Responsibility
De Montfort University
Leicester, UK

Doris Schroeder
Centre for Professional Ethics
University of Central Lancashire
Preston, UK

Rowena Rodrigues
Trilateral Research
London, UK

ISSN 2452-0519   ISSN 2452-0527 (electronic)
SpringerBriefs in Research and Innovation Governance
ISBN 978-3-031-17039-3   ISBN 978-3-031-17040-9 (eBook)
https://doi.org/10.1007/978-3-031-17040-9

This Springer imprint is published by the registered company Springer Nature Switzerland AG
The registered company address is: Gewerbestrasse 11, 6330 Cham, Switzerland

Acknowledgements

This book draws on the work that the authors have done across a range of projects. The primary project that brought us together and demonstrated the need for cases describing ethical issues of AI and ways of addressing them was the EU-funded SHERPA project (2018–2021). All three authors were members of the project, which was led by Bernd Stahl. We would also like to acknowledge the contribution of the members of the SHERPA consortium during the project that informed and inspired the work in this book.

The three authors have been, or still are, active in many other projects that have informed this work directly or indirectly. These projects and the individuals who contributed to them deserve to be acknowledged. They include the EU-funded Human Brain Project, TechEthos and SIENNA, as well as the Responsible-Industry, CONSIDER, TRUST and ETICA projects.

We furthermore acknowledge the support of colleagues in our institutions and organisations, notably the Centre for Computing and Social Responsibility of De Montfort University, the Centre for Professional Ethics at the University of Central Lancashire and Trilateral Research Ltd.

We want to thank Paul Wise for his outstanding editing, Julie Cook for highly insightful comments on the first draft and Jayanthi Krishnamoorthi, Juliana Pitanguy and Toni Milevoj at Springer Nature for overseeing the publishing process very effectively. Thanks to Kostas Iatridis for excellent academic editorial support. We are indebted to Amanda Sharkey for permission to use her exceptionally well-drawn vignette on care robots and the elderly.

Last, but definitely not least, our thanks to three anonymous reviewers whose comments helped us greatly to make changes to the original book ideas.

This research received funding from the European Union's Horizon 2020 Framework Programme for Research and Innovation under Grant Agreements No. 786641 (SHERPA) and No. 945539 (Human Brain Project SGA3).

Contents

Abbreviations

AI HLEG	High-Level Expert Group on AI (EU)
AI	Artificial intelligence
AI-IA	AI impact assessment
ALTAI	Assessment List for Trustworthy AI (EU)
AT + ALP	Attention and Adversarial Logit Pairing
CASR	Campaign Against Sex Robots
CEN	European Committee for Standardization
CENELEC	European Electrotechnical Committee for Standardization
CNIL	Commission Nationale de l'Informatique et des Libertés (France)
COMPAS	Correctional Offender Management Profiling for Alternative Sanctions
DPIA	Data protection impact assessment
DSPO	Detect and suppress the potential outliers
EDPB	European Data Protection Board (EU)
EDPS	European Data Protection Supervisor (EU)
ENISA	European Union Agency for Cybersecurity (EU)
FRA	Fundamental Rights Agency (EU)
GAN	Generative adversarial network
GDPR	General Data Protection Regulation (EU)
GM	Genetically modified
HR	Human resources
ICO	Information Commissioner's Office (UK)
ICT	Information and communication technology
ISPA	International Society of Precision Agriculture
IT	Information technology
NTSB	National Transportation Safety Board (USA)
OECD	Organisation for Economic Co-operation and Development
PLD	Product Liability Directive (EU)
SCF	Seasonal climate forecasting
SDGs	Sustainable Development Goals (UN)
SHGP	Saudi Human Genome Program

UNCTAD	UN Conference on Trade and Development
UNDP	UN Development Programme
UNESCO	UN Educational, Scientific and Cultural Organization
UNSDG	UN Sustainable Development Group
WHO	World Health Organization

Chapter 1
The Ethics of Artificial Intelligence: An Introduction

Abstract This chapter introduces the themes covered by the book. It provides an overview of the concept of artificial intelligence (AI) and some of the technologies that have contributed to the current high level of visibility of AI. It explains why using case studies is a suitable approach to engage a broader audience with an interest in AI ethics. The chapter provides a brief overview of the structure and logic of the book by indicating the content of the cases covered in each section. It concludes by identifying the concept of ethics used in this book and how it is located in the broader discussion of ethics, human rights and regulation of AI.

Keywords Artificial intelligence · Machine learning · Deep learning ethics

The ethical challenges presented by artificial intelligence (AI) are one of the biggest topics of the twenty-first century. The potential benefits of AI are said to be numerous, ranging from operational improvements, such as the reduction of human error (e.g. in medical diagnosis), to the use of robots in hazardous situations (e.g. to secure a nuclear plant after an accident). At the same time, AI raises many ethical concerns, ranging from algorithmic bias and the digital divide to serious health and safety concerns.

The field of AI ethics has boomed into a global enterprise with a wide variety of players. Yet the ethics of artificial intelligence (AI) is nothing new. The concept of AI is almost 70 years old (McCarthy et al. 2006) and ethical concerns about AI have been raised since the middle of the twentieth century (Wiener 1954; Dreyfus 1972; Weizenbaum 1977). The debate has now gained tremendous speed thanks to wider concerns about the use and impact of better algorithms, the growing availability of computing resources and the increasing amounts of data that can be used for analysis (Hall and Pesenti 2017).

These technical developments have favoured specific types of AI, in particular machine learning (Alpaydin 2020; Faggella 2020), of which deep learning is one popular form (see box) (LeCun et al. 2015). The success of these AI approaches led to a rapidly expanding set of uses and applications which frequently resulted in consequences that were deemed ethically problematic, such as unfair or illegal discrimination, exclusion and political interference.

Deep Learning

Deep learning is one of the approaches to machine learning that have led to the remarkable successes of AI in recent years (Bengio et al. 2021). The development of deep learning is a result of the use of artificial neural networks, which are attempts to replicate or simulate brain functions. Natural intelligence arises from parallel networks of neurons that learn by adjusting the strengths of their connections. Deep learning attempts to perform brain-like activities using statistical measures to determine how well a network is performing. Deep learning derives its name from deep neural networks, i.e. networks with many layers. It has been successfully applied to problems ranging from image recognition to natural speech processing. Despite its successes, deep learning has to contend with a range of limitations (Cremer 2021). It is open to debate how much further machine learning based on approaches like deep learning can progress and whether fundamentally different principles might be required, such as the introduction of causality models (Schölkopf et al. 2021).

With new uses of AI, AI ethics has flourished well beyond academia. For instance, the Rome Call for AI Ethics,[1] launched in February 2020, links the Vatican with the UN Food and Agriculture Organization (FAO), Microsoft, IBM and the Italian Ministry of Innovation. Another example is that UNESCO appointed 24 experts from around the world in July 2021 and launched a worldwide online consultation on AI ethics and facilitated dialogue with all UNESCO member states. Media interest is also considerable, although some academics consider the treatment of AI ethics by the media as "shallow" (Ouchchy et al. 2020).

One of the big problems that AI ethics and ethicists might face is the opaqueness of what is actually happening in AI, given that a good grasp of an activity itself is very helpful in determining its ethical issues.

[I]t is not the role nor to be expected of an AI Ethicist to be able to program the systems themselves. Instead, a strong understanding of aspects such as the difference between supervised and unsupervised learning, what it means to label a dataset, how consent of the user is obtained – essentially, how a system is designed, developed, and deployed – is necessary. In other words, an AI Ethicist must comprehend enough to be able to apprehend the instances in which key ethical questions must be answered (Gambelin 2021).

There is thus an expectation that AI ethicists are familiar with the technology, yet "[n]o one really knows how the most advanced algorithms do what they do" (Knight 2017), including AI developers themselves.

Despite this opacity of AI in its current forms, it is important to reflect on and discuss which ethical issues can arise due to its development and use. The approach to AI ethics we have chosen here is to use case studies, as "[r]eal experiences in AI ethics present … nuanced examples" (Brusseau 2021) for discussion, learning and

---

[1] https://www.romecall.org/.

analysis. This approach will enable us to illustrate the main ethical challenges of AI, often with reference to human rights (Franks 2017).

Case studies are a proven method for increasing insights into theoretical concepts by illustrating them through real-world situations (Escartín et al. 2015). They also increase student participation and enhance the learning experience (ibid) and are therefore well-suited to teaching (Yin 2003).

We have therefore chosen the case study method for this book. We selected the most significant or pertinent ethical issues that are currently discussed in the context of AI (based on and updated from Andreou et al. 2019 and other sources) and dedicated one chapter to each of them.

The structure of each chapter is as follows. First, we introduce short real-life case vignettes to give an overview of a particular ethical issue. Second, we present a narrative assessment of the vignettes and the broader context. Third, we suggest ways in which these ethical issues could be addressed. This often takes the form of an overview of the tools available to reduce the ethical risks of the particular case; for instance, a case study of algorithmic bias leading to discrimination will be accompanied by an explanation of the purpose and scope of AI impact assessments. Where tools are not appropriate, as human decisions need to be made based on ethical reasoning (e.g. in the case of sex robots), we provide a synthesis of different argument strategies. Our focus is on *real-life* scenarios, most of which have already been published by the media or research outlets. Below we present a short overview of the cases.

Unfair and Illegal Discrimination (Chap. 2)

The first vignette deals with the automated shortlisting of job candidates by an AI tool trained with CVs (résumés) from the previous ten years. Notwithstanding efforts to address early difficulties with gender bias, the company eventually abandoned the approach as it was not compatible with their commitment to workplace diversity and equality.

The second vignette describes how parole was denied to a prisoner with a model rehabilitation record based on the risk-to-society predictions of an AI system. It became clear that subjective personal views given by prison guards, who may have been influenced by racial prejudices, led to an unreasonably high risk score.

The third vignette tells the story of an engineering student of Asian descent whose passport photo was rejected by New Zealand government systems because his eyes were allegedly closed. This was an ethnicity-based error in passport photo recognition, which was also made by similar systems elsewhere, affecting, for example, dark-skinned women in the UK.

Privacy (Chap. 3)

The first vignette is about the Chinese social credit scoring system, which uses a large number of data points to calculate a score of citizens' trustworthiness. High scores lead to the allocation of benefits, whereas low scores can result in the withdrawal of services.

The second vignette covers the Saudi Human Genome Program, with predicted benefits in the form of medical breakthroughs versus genetic privacy concerns.

Surveillance Capitalism (Chap. 4)

The first vignette deals with photo harvesting from services such as Instagram, LinkedIn and YouTube in contravention of what users of these services were likely to expect or have agreed to. The relevant AI software company, which specialises in facial recognition software, reportedly holds ten billion facial images from around the world.

The second vignette is about a data leak from a provider of health tracking services, which made the health data of 61 million people publicly available.

The third vignette summarises Italian legal proceedings against Facebook for misleading its users by not explaining to them in a timely and adequate manner, during the activation of their account, that data would be collected with commercial intent.

Manipulation (Chap. 5)

The first vignette covers the Facebook and Cambridge Analytica scandal, which allowed Cambridge Analytica to harvest 50 million Facebook profiles, enabling the delivery of personalised messages to the profile holders and a wider analysis of voter behaviour in the run-up to the 2016 US presidential election and the Brexit referendum in the same year.

The second vignette shows how research is used to push commercial products to potential buyers at specifically determined vulnerable moments, e.g. beauty products being promoted at times when recipients of online commercials are likely to feel least attractive.

Right to Life, Liberty and Security of Person (Chap. 6)

The first vignette is about the well-known crash of a Tesla self-driving car, killing the person inside.

The second vignette summarises the security vulnerabilities of smart home hubs, which can lead to man-in-the-middle attacks, a type of cyberattack in which the security of a system is compromised, allowing an attacker to eavesdrop on confidential information.

The third vignette deals with adversarial attacks in medical diagnosis, in which an AI-trained system could be fooled to the extent of almost 70% with fake images.

Dignity (Chap. 7)

The first vignette describes the case of an employee who was wrongly dismissed and escorted off his company's premises by security guards, with implications for his dignity. The dismissal decision was based on opaque decision-making by an AI tool, communicated by an automatic system.

The second vignette covers sex robots, in particular whether they are an affront to the dignity of women and female children.

Similarly, the third vignette asks whether care robots are an affront to the dignity of elderly people.

AI for Good and the UN's Sustainable Development Goals (Chap. 8)

The first vignette shows how seasonal climate forecasting in resource-limited settings has led to the denial of credits for poor farmers in Zimbabwe and Brazil and the accelerated the layoff of workers in the fishing industry in Peru.

The second vignette deals with a research team from a high-income country requesting vast amounts of mobile phone data from users in Sierra Leone, Guinea and Liberia to track population movements during the Ebola crisis. Commentators argued that the time spent negotiating the request with seriously under-resourced governance structures should have been used to handle the escalating Ebola crisis.

This is a book of AI ethics case studies and not a philosophical book on ethics. We nevertheless need to be clear about our use of the term "ethics". We use the concept of ethics cognisant of the venerable tradition of ethical discussion and of key positions such as those based on an evaluation of the duty of an ethical agent (Kant 1788, 1797), the consequences of an action (Bentham 1789; Mill 1861), the character of the agent (Aristotle 2000) and the keen observation of potential biases in one's own position, for instance through using an ethics of care (Held 2005). We slightly favour a Kantian position in several chapters, but use and acknowledge others. We recognize that there are many other ethical traditions beyond the dominant European ones mentioned here, and we welcome debate about how these may help us understand further aspects of ethics and technology. We thus use the term "ethics" in a pluralistic sense.

This approach is pluralistic because it is open to interpretations from the perspective of the main ethical theories as well as other theoretical positions, including more recent attempts to develop ethical theories that are geared more specifically to novel technologies, such as disclosive ethics (Brey 2000), computer ethics (Bynum 2001), information ethics (Floridi 1999) and human flourishing (Stahl 2021).

Our pluralistic reading of the ethics of AI is consistent with much of the relevant literature. A predominant approach to AI ethics is the development of guidelines (Jobin et al. 2019), most of which are based on mid-level ethical principles typically developed from the principles of biomedical ethics (Childress and Beauchamp 1979). This is also the approach adopted by the European Union's High-Level Expert Group on AI (AI HLEG 2019). The HLEG's intervention has been influential, as it has had a great impact on the discussion in Europe, which is where we are physically located and which is the origin of the funding for our work (see Acknowledgements). However, there has been significant criticism of the approach to AI ethics based on ethical principles and guidelines (Mittelstadt 2019; Rességuier and Rodrigues 2020). One key concern is that it remains far from the application and does not explain how AI ethics can be put into practice. With the case-study-based approach presented in this book, we aim to overcome this point of criticism, enhance ethical reflection and demonstrate possible practical interventions.

We invite the reader to critically accompany us on our journey through cases of AI ethics. We also ask the reader to think beyond the cases presented here and ask fundamental questions, such as whether and to what degree the issues discussed here are typical or exclusively relevant to AI and whether one can expect them to be resolved.

Overall, AI is an example of a current and dynamically developing technology. An important question is therefore whether we can keep reflecting and learn anything from the discussion of AI ethics that can be applied to future generations of technologies to ensure that humanity benefits from technological progress and development and has ways to deal with the downsides of technology.

References

AI HLEG (2019) Ethics guidelines for trustworthy AI. High-level expert group on artificial intelligence. European Commission, Brussels. https://ec.europa.eu/newsroom/dae/document.cfm?doc_id=60419. Accessed 25 Sept 2020

Alpaydin E (2020) Introduction to machine learning. The MIT Press, Cambridge

Andreou A, Laulhe Shaelou S, Schroeder D (2019) D1.5 Current human rights frameworks. De Montfort University. Online resource. https://doi.org/10.21253/DMU.8181827.v3

Aristotle (2000) Nicomachean ethics (trans: Crisp R). Cambridge University Press, Cambridge

Bengio Y, Lecun Y, Hinton G (2021) Deep learning for AI. Commun ACM 64:58–65. https://doi.org/10.1145/3448250

Bentham J (1789) An introduction to the principles of morals and legislation. Dover Publications, Mineola

Brey P (2000) Disclosive computer ethics. SIGCAS Comput Soc 30(4):10–16. https://doi.org/10.1145/572260.572264

Brusseau J (2021) Using edge cases to disentangle fairness and solidarity in AI ethics. AI Ethics. https://doi.org/10.1007/s43681-021-00090-z

Bynum TW (2001) Computer ethics: its birth and its future. Ethics Inf Technol 3:109–112. https://doi.org/10.1023/A:1011893925319

Childress JF, Beauchamp TL (1979) Principles of biomedical ethics. Oxford University Press, New York

Cremer CZ (2021) Deep limitations? Examining expert disagreement over deep learning. Prog Artif Intell 10:449–464. https://doi.org/10.1007/s13748-021-00239-1

Dreyfus HL (1972) What computers can't do: a critique of artificial reason. Harper & Row, New York

Escartín J, Saldaña O, Martín-Peña J et al (2015) The impact of writing case studies: benefits for students' success and well-being. Procedia Soc Behav Sci 196:47–51. https://doi.org/10.1016/j.sbspro.2015.07.009

Faggella D (2020) Everyday examples of artificial intelligence and machine learning. Emerj, Boston. https://emerj.com/ai-sector-overviews/everyday-examples-of-ai/. Accessed 23 Sept 2020

Floridi L (1999) Information ethics: on the philosophical foundation of computer ethics. Ethics Inf Technol 1:33–52. https://doi.org/10.1023/A:1010018611096

Franks B (2017) The dilemma of unexplainable artificial intelligence. Datafloq, 25 July. https://datafloq.com/read/dilemma-unexplainable-artificial-intelligence/. Accessed 18 May 2022

Gambelin O (2021) Brave: what it means to be an AI ethicist. AI Ethics 1:87–91. https://doi.org/10.1007/s43681-020-00020-5

Hall W, Pesenti J (2017) Growing the artificial intelligence industry in the UK. Department for Digital, Culture, Media & Sport and Department for Business, Energy & Industrial Strategy, London

Held V (2005) The ethics of care: personal, political, and global. Oxford University Press, New York

Jobin A, Ienca M, Vayena E (2019) The global landscape of AI ethics guidelines. Nat Mach Intell 1:389–399. https://doi.org/10.1038/s42256-019-0088-2

Kant I (1788) Kritik der praktischen Vernunft. Reclam, Ditzingen

Kant I (1797) Grundlegung zur Metaphysik der Sitten. Reclam, Ditzingen

Knight W (2017) The dark secret at the heart of AI. MIT Technology Review, 11 Apr. https://www.technologyreview.com/2017/04/11/5113/the-dark-secret-at-the-heart-of-ai/. Accessed 18 May 2022

LeCun Y, Bengio Y, Hinton G (2015) Deep learning. Nature 521:436–444. https://doi.org/10.1038/nature14539

McCarthy J, Minsky ML, Rochester N, Shannon CE (2006) A proposal for the Dartmouth summer research project on artificial intelligence. AI Mag 27:12–14. https://doi.org/10.1609/aimag.v27i4.1904

Mill JS (1861) Utilitarianism, 2nd revised edn. Hackett Publishing Co, Indianapolis

Mittelstadt B (2019) Principles alone cannot guarantee ethical AI. Nat Mach Intell 1:501–507. https://doi.org/10.1038/s42256-019-0114-4

Ouchchy L, Coin A, Dubljević V (2020) AI in the headlines: the portrayal of the ethical issues of artificial intelligence in the media. AI & Soc. https://doi.org/10.1007/s00146-020-00965-5

Rességuier A, Rodrigues R (2020) AI ethics should not remain toothless! A call to bring back the teeth of ethics. Big Data Soc 7:2053951720942541. https://doi.org/10.1177/2053951720942541

Schölkopf B, Locatello F, Bauer S et al (2021) Toward causal representation learning. Proc IEEE 109(5):612–634. https://doi.org/10.1109/JPROC.2021.3058954

Stahl BC (2021) Artificial intelligence for a better future: an ecosystem perspective on the ethics of AI and emerging digital technologies. Springer Nature Switzerland AG, Cham. https://doi.org/10.1007/978-3-030-69978-9

UNESCO (2021) AI ethics: another step closer to the adoption of UNESCO's recommendation. UNESCO, Paris. Press release, 2 July. https://en.unesco.org/news/ai-ethics-another-step-closer-adoption-unescos-recommendation-0. Accessed 18 May 2022

Weizenbaum J (1977) Computer power and human reason: from judgement to calculation, new edn. W.H. Freeman & Co Ltd., New York

Wiener N (1954) The human use of human beings. Doubleday, New York

Yin RK (2003) Applications of case study research, 2nd edn. Sage Publications, Thousand Oaks

Chapter 2
Unfair and Illegal Discrimination

Abstract There is much debate about the ways in which artificial intelligence (AI) systems can include and perpetuate biases and lead to unfair and often illegal discrimination against individuals on the basis of protected characteristics, such as age, race, gender and disability. This chapter describes three cases of such discrimination. It starts with an account of the use of AI in hiring decisions that led to discrimination based on gender. The second case explores the way in which AI can lead to discrimination when applied in law enforcement. The final example looks at implications of bias in the detection of skin colour. The chapter then discusses why these cases are considered to be ethical issues and how this ethics debate relates to well-established legislation around discrimination. The chapter proposes two ways of raising awareness of possible discriminatory characteristics of AI systems and ways of dealing with them: AI impact assessments and ethics by design.

Keywords Discrimination · Bias · Gender · Race · Classification · Law enforcement · Predictive policing · AI impact assessment · Ethics by design

2.1 Introduction

Concern at discrimination is probably the most widely discussed and recognised ethical issue linked to artificial intelligence (AI) (Access Now 2018; Latonero 2018; Muller 2020). In many cases an AI system analyses existing data which was collected for purposes other than the ones that the AI system is pursuing and therefore typically does so without paying attention to properties of the data that may facilitate unfair discrimination when used by the AI system. Analysis of the data using AI reveals underlying patterns that are then embedded in the AI model used for decision-making. In these cases, which include our examples of gender bias in staff recruitment and predictive policing that disadvantages segments of the population, the system perpetuates existing biases and reproduces prior practices of discrimination.

In some cases, discrimination occurs through other mechanisms, for example when a system is exposed to real-world data that is fundamentally different from the data it was trained on and cannot process the data correctly. Our case of systems that misclassify people from ethnic groups that are not part of the training data falls into this category. In this case the system works in a way that is technically correct, but the outputs are incorrect, due to a lack of correspondence between the AI model and the input data.

These examples of AI-enabled discrimination have in common that they violate a human right (see box) that individuals should not be discriminated against. That is why these systems deserve attention and are the subject of this chapter.

Universal Declaration of Human Rights, Article 7

"All are equal before the law and are entitled without any discrimination to equal protection of the law. All are entitled to equal protection against any discrimination in violation of this Declaration and against any incitement to such discrimination." (UN 1948)

2.2   Cases of AI-Enabled Discrimination

2.2.1   Case 1: Gender Bias in Recruitment Tools

Recruiting new members of staff is an important task for an organisation, given that human resources are often considered the most valuable assets a company can have. At the same time, recruitment can be time- and resource-intensive. It requires organisations to scrutinise job applications and CVs, which are often non-standardised, complex documents, and to make decisions on shortlisting and appointments on the basis of this data. It is therefore not surprising that recruitment was an early candidate for automation by machine learning. One of the most high-profile examples of AI use for recruitment is an endeavour by Amazon to automate the candidate selection process.

In 2014, Amazon started to develop and use AI programs to mechanise highly time-intensive human resources (HR) work, namely the shortlisting of applicants for jobs. Amazon "literally wanted it to be an engine where I'm going to give you 100 résumés, it will spit out the top five, and we'll hire those" (Reuters 2018). The AI tool was trained on CVs submitted over an earlier ten-year period and the related staff appointments. Following this training, the AI tool discarded the applications of female applicants, even where no direct references to applicants' gender were provided. Given the predominance of successful male applicants in the training sample, Amazon found that the system penalised language such as "women's chess club captain" for not matching closely enough the successful male job applicants of the past. While developers tried to modify the system to avoid gender bias, Amazon abandoned its use in the recruitment process in 2015 as a company "committed to workplace diversity and equality" (ibid).

At first this approach seemed promising, as HR departments have ample training data in the form of past applications. A machine learning system can thus be trained to distinguish between successful and unsuccessful past applications and identify features of applications that are predictors of success. This is exactly what Amazon did. The result was that the AI systematically discriminated against women.

When it became clear that women were being disadvantaged by recruitment based on AI, ways were sought to fix the problem. The presumptive reason for the outcome was that there were few women in the training sample, maybe because the tech sector is traditionally male dominated, or maybe reflecting biases in the recruitment system overall. It turned out, however, that even removing direct identifiers of sex and gender did not level the playing field, as the AI found proxy variables that still pointed to gender, such as place of study (e.g., all-female college) and feminised hobbies.

> AI systems are only as good as the data they're trained on and the humans that build them. If a résumé-screening machine-learning tool is trained on historical data, such as résumés collected from a company's previously hired candidates, the system will inherit both the conscious and unconscious preferences of the hiring managers who made those selections (Heilweil 2019).

In the case of Amazon this eventually led to the company's abandoning the use of AI for hiring, as explained in the case description. However, the fundamental challenge of matching large numbers of candidates for recruitment with large numbers of open positions on the basis of complex and changing selection criteria remains. For instance, Vodafone is reported to have used AI systems to analyse over 100,000 graduate applications for 1,000 jobs (Kaur 2021). Since 2019, the COVID-19 pandemic has accelerated the use of AI recruitment, with predictions that 16% of HR recruitment jobs will have disappeared by 2029 (ibid).

AI can also, it is claimed, be used as a tool for measuring psychological, emotional and personality features during video interviews (Heilweil 2019). Online interviews have become the norm under COVID-19 lockdowns, and this trend seems set to continue, so the use of AI technology in these contexts may increase. However, tools that interpret facial features may manifest limitations similar to those of recruitment AI, although their impact is not as widely publicised as that of the Amazon case. This means that sustained ethical alertness is required when it comes to preventing violations of the human right to non-discrimination. Or, as a human rights commentator has noted, the problem of "garbage in, garbage out" (Lentz 2021) has to be solved before HR departments can use AI in an ethical manner to substitute human for machine decision-making.

2.2.2 Case 2: Discriminatory Use of AI in Law Enforcement and Predictive Policing

Glenn Rodríguez had been arrested at the age of 16 for his role in the armed robbery of a car dealership, which left one employee dead. In 2016, 25 years later, he applied to the parole board of the Eastern Correctional Facility in upstate New York for early release. He had a model rehabilitation record at the time (Wexler 2017b). Parole was denied. The justification given by the board was that an AI system called COMPAS had predicted him to be "high risk" and the board "concluded that ... release to supervision is not compatible with the welfare of society" (Wexler 2017a). The parole board had no knowledge of how the COMPAS risk score was calculated, as the company that had developed the system considered their algorithm a trade secret (ibid). Through cross-referencing with other inmates' scores, Rodríguez found out that the reason for his high-risk score was a subjective personal view given by prison guards, who may have been influenced by racial prejudices. In the end, he was released early. However, "had he been able to examine and contest the logic of the COMPAS system to prove that its score gave a distorted picture of his life, he might have gone home much earlier" (Wexler 2017b)

Rodríguez's case is an example of the discriminatory use of AI in criminal justice, which also includes prominent AI applications for the purposes of predictive policing. "Predictive policing makes use of information technology, data, and analytical techniques in order to identify likely places and times of future crimes or individuals at high risk of [re-]offending or becoming victims of crime." (Mugari and Obioha 2021: 1). The idea behind predictive policing is that existing law enforcement data can improve the targeting of policing interventions. Police resources are limited and it would be desirable to focus them where they are most likely to make a difference, that is, to disrupt or prevent crime or, once crime has been committed, to protect victims, arrest offenders etc. Predictive policing uses past crime data to detect patterns suitable for extrapolation into the future, thereby, one hopes, helping police to identify locations and times when crime is most likely to occur. This is where resources are then deployed.

These ideas sound plausible and are already implemented in many jurisdictions. The most high-profile cases are from the US, where police have been developing and using predictive policing tools in Chicago, Los Angeles, New Orleans and New York since as far back as 2012 (McCarthy 2019). In the UK, research by an NGO showed that "at least 14 UK police forces have used or intend to use ... computer algorithms to predict where crime will be committed and by whom" (Liberty n.d.). It is also known that China, Denmark, Germany, India, the Netherlands, and Japan are testing and possibly deploying predictive policing tools (McCarthy 2019).

While the idea of helping the police do their job better, and possibly at reduced cost, will be welcomed by many, the practice of predictive policing has turned out to be ethically problematic. The use of past crime data means that historical patterns are reproduced, and this may become a self-fulfilling prophecy.

For example, areas that historically have high crime rates tend to be those that have lower levels of wealth and educational attainment among the population, as well as higher percentages of migrants or stateless people. Using predictive policing tools means that people who live in deprived areas are singled out for additional police attention, whether they have anything to do with perpetrating any crimes or not. Using algorithmic systems to support policing work has the potential to exacerbate already entrenched discrimination. It is worth pointing out, however, that given awareness of the issue, it is also conceivable that such systems could explicitly screen police activity for bias and help alleviate the problem. The AI systems used for predictive policing and law enforcement could be used to extract and visualise crime data that would make more obvious whether and how crime statistics are skewed in ways that might be linked to ethnic or racial characteristics. This, in turn, would provide a good starting point for a more detailed analysis of the mechanisms that contribute to such developments.

This problem of possible discrimination in relation to specific geographical areas can also occur in relation to individuals. Automated biometric recognition can be used in police cameras, providing police officers with automated risk scores for people they interact with. This then disadvantages people with prior convictions or a past history of interaction with the police, which again tends to over-represent disadvantaged communities, notably those from ethnic minorities. The same logic applies further down the law enforcement chain, when the analysis of data from offenders is used to predict their personal likelihood of reoffending. When the AI tool which informed the decision to hold Glenn Rodríguez in prison for longer than necessary was later examined, it was found that "a disproportionate number of black defendants were 'false positives': they were classified by COMPAS as high risk but subsequently not charged with another crime." (Courtland 2018).

2.2.3 Case 3: Discrimination on the Basis of Skin Colour

In 2016, a 22-year-old engineering student from New Zealand had his passport photo rejected by the systems of the New Zealand department of internal affairs because his eyes were allegedly closed. The student was of Asian descent and his eyes were open. The automatic photo recognition tool declared the photo invalid and the student could not renew his passport. He later told the press very graciously: "No hard feelings on my part, I've always had very small eyes and facial recognition technology is relatively new and unsophisticated" (Reuters 2016). Similar cases of ethnicity-based errors by passport photo recognition tools have affected dark-skinned women in the UK. "Photos of women with the darkest skin were four times more likely to be graded poor quality, than women with the lightest skin" (Ahmed 2020). For instance, a black student's photo was declared unsuitable as her mouth was allegedly open, which it in fact was not (ibid).

Zou and Schiebinger (2018) have explained how such discriminatory bias can occur. As noted earlier, one of the main reasons for discriminatory AI tools is the training sets used.

> Deep neural networks for image classification … are often trained on ImageNet … More than 45% of ImageNet data, which fuels research in computer vision, comes from the United States, home to only 4% of the world's population.

Hence, some groups are heavily over-represented in training sets while others are under-represented, leading to the perpetuation of ethnicity-based discrimination.

2.3   Ethical Questions Concerning AI-Enabled Discrimination

The reproduction of biases and resulting discrimination are among the most prominent ethical concerns about AI (Veale and Binns 2017; Access Now Policy Team 2018). Bias has been described as the "one of the biggest risks associated with AI" (PwC 2019: 13).

The term "discrimination" has at least two distinct meanings, which differ significantly in terms of an ethical analysis (Cambridge Dictionary n.d.). On one hand "discrimination" means the ability to judge phenomena and distinguish between them in a reasonable manner. In this sense, the term has synonyms like "distinction" and "differentiation". For instance, it is a good evolutionary trait for humans to have the ability to distinguish malaria-carrying mosquitoes from flies. The other more widespread contemporary meaning of the term focuses on the unjust or prejudicial application of distinctions made between people, in particular on the basis of their race, sex, age or disability. The former meaning can be ethically neutral, whereas the latter is generally acknowledged to be a significant ethical problem, hence article 7 of the Universal Declaration of Human Rights (see box above). When we use the term "discrimination" in this discussion, we are talking about the ethically relevant type, which is also often illegal.

However, being able to distinguish between phenomena is one of the strengths of AI. Machine-learning algorithms are specifically trained to distinguish between classes of phenomena, and their success in doing so is the main reason for the current emphasis on AI use in a wide field of applications.

AI systems have become increasingly adept at drawing distinctions, at first between pictures of cats and pictures of dogs, which provided the basis for their use in more socially relevant fields, such as medical pathology, where they can distinguish images of cancer cells from those of healthy tissue, or in the business world, where they can distinguish fraudulent insurance claims from genuine ones. The problem is not identifying differences in the broad sense but discrimination on the basis of those particular characteristics.

Unfair/illegal discrimination is a widespread characteristic of many social interactions independent of AI use. While there is broad agreement that job offers should not depend on an applicant's gender, and that judicial or law enforcement decisions should not depend on a person's ethnicity, it is also clear that they often do, reflecting ingrained systemic injustices. An AI system that is trained on historical data that includes data from processes that structurally discriminated against people will replicate that discrimination. As our case studies have shown, these underlying patterns in the data are difficult to eradicate. Attempts to address such problems by providing more inclusive data may offer avenues for overcoming them. However, there are many cases where no alternative relevant datasets exist. In such cases, which include law enforcement and criminal justice applications, the attempt to modify the data to reduce or eliminate underlying biases may inadvertently introduce new challenges.

However, there are cases where the problem is not so much that no unbiased datasets exist but that the possibility of introducing biases through a poor choice of training data is not sufficiently taken into account. An example is unfair/illegal discrimination arising from poor systems design through a poor choice of training data. Our third case study points in this direction. When the images of 4% of the world population constitute 45% of the images used in AI system design (Zou and Schiebinger 2018), it is reasonable to foresee unfair/illegal discrimination in the results.

This type of discrimination will typically arise when a machine-learning system is trained on data that does not fully represent the population that the system is meant to be applied to. Skin colour is an obvious example, where models based on data from one ethnic group do not work properly when applied to a different group. Such cases are similar to the earlier ones (Amazon HR and parole) in that there is a pre-existing bias in the original data used to train the model. The difference between the two types of discrimination is the source of the bias in the training data. In the first two cases the biases were introduced by the systems involved in creating the data, i.e. in recruitment processes and law enforcement, where women and racial minorities were disadvantaged by past recruiters and past parole boards that had applied structurally sexist or racist perspectives. In the case of discrimination based on skin colour, the bias was introduced by a failure to select comprehensive datasets that included representation from all user communities. This difference is subtle and not always clear-cut. It may be important, however, in that ways of identifying and rectifying particular problems may differ significantly.

Discrimination against people on the basis of gender, race, age etc. is not only an ethical issue; in many jurisdictions such discrimination is also illegal. In the UK, for example, the Equality Act (2010) defines nine protected characteristics: age, disability, gender reassignment, marriage and civil partnership, pregnancy and maternity, race, religion or belief, sex, and sexual orientation. Discrimination in the workplace and in wider society based on these protected characteristics is prohibited.

The legal codification of the prohibition of such discrimination points to a strong societal consensus that such discrimination is to be avoided. It raises difficult questions, however, with regard to unfair discrimination that is based on characteristics other than the legally protected ones. It is conceivable that a system would identify patterns on the basis of other variables that we may not yet even be aware of. Individuals could then be categorised in ways that are detrimental to them. This might not involve protected characteristics, but could still be perceived as unfair discrimination.

Another example of a problematic variable is social class. It is well established that class is an important variable that determines not just individual life chances, but also the collective treatment of groups. Marx's (2017) dictum that the history of all existing society is the history of class struggles exemplifies this position. Discrimination can happen because of a particular characteristic, such as gender, race or disability, but it often happens where individuals combine several of these characteristics that individually can lead to discrimination and, when taken together, exacerbate the discriminatory effect. The term "intersectionality" is sometimes used to indicate this phenomenon (Collins and Bilge 2020). Intersectionality has been recognised as a concern that needs to be considered in various aspects of information technology (IT), not only AI (Fothergill et al. 2019; Zheng and Walsham 2021). It points to the fact that the exact causes of discrimination will in practice often be difficult to identify, which raises questions about the mechanisms of unfair/illegal discrimination as well as ways of addressing them. If the person who is discriminated against is a black, disabled, working-class woman, then it may be impossible to determine which characteristic led to the discrimination, and whether the discrimination was based on protected characteristics and thus illegal.

Hence, unfair/illegal discrimination is not a simple matter. Discrimination based on protected characteristics is deemed to be ethically unacceptable in most democratic states and therefore also typically illegal. But this does not mean that there is no discrimination in social reality, nor should we take it as given that the nature of these protected characteristics will remain constant or that discrimination based on gender, age, race etc. are the only forms of unfair discrimination.

2.4   Responses to Unfair/Illegal Discrimination

With unfair/illegal discrimination recognised as a key ethical problem related to AI and machine learning, there is no shortage of attempts to address and mitigate it. These range from the technical level, where attempts are made to better understand whether training data contains biases that lead to discrimination, to legislative processes where existing anti-discrimination policies are refocused on novel technologies.

One prominent field of research with significant implications regarding unfair/illegal discrimination is that of explainable AI (Holzinger et al. 2017; Gunning et al. 2019). There are many approaches to explainable AI, but what they have in common is an attempt to render the opaque nature of the transformation from input variables to output variables easier to understand. The logic is that an ability to understand *how* an AI system came to a classification of a particular observation would allow the determination of whether that classification is discriminatory and, as a result, could be challenged. If AI is fully explainable, then it should be easy to see whether gender (sexism) determines employment offers, or whether racism has consequences for law enforcement practices.

While this approach is plausible, it runs into technical and social limits. The technical limits include the fact that machine learning models include large numbers of variables and by their very nature are not easy to understand. If it were possible to reduce them to simple tests of specific variables, then machine learning would not be needed in the first place. However, it might be possible for explainable AI to find ways of testing whether an AI system makes use of protected characteristics and to correct for this (Mittelstadt 2019). Hence, rather than humans making these assessments, another or the same AI system would do so.

When thinking about ways of addressing the role that AI plays in unfair/illegal discrimination, it helps to keep in mind that such discrimination is pervasive in many social processes. Real-life data used for training purposes will often include cases of unfair discrimination and thus lead to their reproduction. Removing traces of structural discrimination from training data, for example by removing data referring to protected characteristics, may not work or may reduce the value of the data for training purposes. The importance of data quality to the trustworthiness of the outcomes of an AI system is widely recognised. The European Commission's proposal for regulating AI, for example, stipulates that "training, validation and testing data sets shall be relevant, representative, free of errors and complete" (European Commission 2021: art. 10(3)). It is not clear, however whether such data quality requirements can possibly be met with real-life data.

Two suggestions on how to address unfair/illegal discrimination (Stahl 2021) will be highlighted here: AI impact assessments and ethics by design.

2.4.1 AI Impact Assessment

The idea of an AI impact assessment is based on the insights derived from many other types of impact assessment, such as social impact assessment (Becker 2001; Becker and Vanclay 2003; Hartley and Wood 2005) and human rights impact assessment (Microsoft and Article One 2018). In general terms, impact assessment aims to come to a better understanding of the possible and likely issues that can arise in a particular field, and use this understanding to prepare mitigation measures. There are several examples of impact assessment that focus on information technologies and topics of relevance to AI, such as privacy/data protection impact assessment (CNIL 2015; Ivanova 2020), ICT ethics impact assessment (Wright 2011) and ethics impact assessment for research and innovation (CEN-CENELEC 2017). The idea of applying impact assessments specifically to AI and using them to get an early warning of possible ethical issues is therefore plausible. This has led to a number of calls for such specific impact assessments for AI by bodies such as the European Data Protection Supervisor (EDPS 2020), UNESCO (2020), the European Fundamental Rights Agency (FRA 2020) and the UK AI council (2021).

The discussion of what such an AI impact assessment should look like in detail is ongoing, but several proposals are available. Examples include the assessment list for trustworthy AI of the European Commission's High-Level Expert Group on Artificial Intelligence (AI HLEG 2020), the AI Now Institute's algorithmic impact assessment (Reisman et al. 2018), the IEEE's recommended practice for assessing the impact of autonomous and intelligent systems on human wellbeing (IEEE 2020) and the ECP Platform's AI impact assessment (ECP 2019).

The idea common to these AI impact assessments is that they provide a structure for thinking about aspects that are likely to raise concerns at a later stage. They highlight such issues and often propose processes to be put in place to address them. In a narrow sense they can be seen as an aspect of risk management. More broadly they can be interpreted as a proactive engagement that typically includes stakeholder consultation to ensure that likely and foreseeable problems do not arise. Bias and unfair/illegal discrimination figure strongly among these foreseeable problems.

The impact assessment aims to ascertain that appropriate mechanisms for dealing with potential sources of bias and unfair discrimination are flagged early and considered by those designing AI systems. The AI HLEG (2020) assessment, for example, asks whether strategies for avoiding biases are in place, how the diversity and representativeness of end users is considered, whether AI designers and developers have benefitted from education and awareness initiatives to sensitise them to the problem, how such issues can be reported and whether a consistent use of the terminology pertaining to fairness is ensured.

An AI impact assessment is therefore likely to be a good way of raising awareness of the possibility and likelihood that an AI system may raise concerns about unfair/illegal discrimination, and of which form this discrimination might take. However, it typically does not go far in providing a pathway towards addressing such discrimination, which is the ambition of ethics by design.

2.4.2 Ethics by Design

Ethics by design for AI has been developed in line with previous discussions of value-sensitive design (Friedman et al. 2008; van den Hoven 2013). The underlying idea is that an explicit consideration of shared values during the design and development process of a project or technology will be conducive to the embedding of such a value in the technology and its eventual use. The concept has been prominently adopted for particular values, for example in the area of privacy by design (ICO 2008) or security by design (Cavoukian 2017).

A key premise of value-sensitive design is that technology is not a value-neutral tool that can be used for any purposes; design decisions influence the way in which a technology can be used and what consequences such use will have. This idea may be most easily exemplified using the value of security by design. Cybersecurity is generally recognised as an important concern that requires continuous vigilance from individuals, organisations and society. It is also well recognised that some systems are easier to protect from malicious interventions than others. One distinguishing factor between more secure and less secure systems is that secure systems tend to be built with security considerations integrated into the earliest stages of systems design. Highlighting the importance of security, for example in the systems requirement specifications, makes it more likely that the subsequent steps of systems development will be sensitive to the relevance of security and ensure that the system overall contains features that support security. Value-sensitive design is predicated on the assumption that a similar logic can be followed for all sorts of values.

The concept of ethics by design was developed by Philip Brey and his collaborators (Brey and Dainow 2020) with a particular view to embedding *ethical* values in the design and development of AI and related technologies. This approach starts by highlighting the values that are likely to be affected by a particular technology. Brey and Dainow (2020) take their point of departure from the AI HLEG (2019) and identify the following values as relevant: human agency, privacy and data governance, fairness, wellbeing, accountability and oversight, and transparency. The value of fairness is key to addressing questions of bias and unfair/illegal discrimination.

Where ethics by design goes beyond an ex-ante impact assessment is where it specifically proposes ways of integrating attention to the relevant values into the design process. For this purpose, Brey and Dainow (2020) look at the way in which software is designed. Starting with a high-level overview, they distinguish different design phases and translate the ethical values into specific objectives and require-ments that can then be fed into the development process. They also propose ways in which this can be achieved in the context of agile development methodologies. This explicit link between ethical concerns and systems development methodologies is a key conceptual innovation of ethics by design. Systems development methodologies are among the foundations of computer science. They aim to ensure that systems can be built according to specifications and perform as expected. The history of computer science has seen the emergence of numerous design methodologies. What Brey and his colleagues have done is to identify universal components that most systems devel-opment methodologies share (e.g. objectives specification, requirements elicitation, coding, testing) and to provide guidance on how ethical values can be integrated and reflected in these steps.

This method has only recently been proposed and has not yet been evaluated. It nevertheless seems to offer an avenue for the practical implementation of ethical values, including the avoidance of unfair/illegal discrimination in AI systems. In light of the pervasive nature of unfair/illegal discrimination in most areas of society one can safely say that all AI systems need to be built and used in ways that recognise the possibility of discrimination. Failure to take this possibility into account means that the status quo will be reproduced using AI, which will often be neither ethical nor legal.

2.5 Key Insights

Unfair/illegal discrimination is not a new problem, nor one that is confined to technology. However, AI systems have the proven potential to exacerbate and perpetuate
it. A key problem in addressing and possibly overcoming unfair/illegal discrimination is that it is pervasive and often hidden from sight. High-profile examples of such
discrimination on the basis of gender and race have highlighted the problem, as in
our case studies. But unfair/illegal discrimination cannot be addressed by looking
at technology alone. The broader societal questions of discrimination need to be
considered.

One should also not underestimate the potential for AI to be used as a tool to
identify cases of unfair/illegal discrimination. The ability of AI to recognise patterns
and process large amounts of data means that AI may also be used to demonstrate
where discrimination is occurring.

It is too early to evaluate whether – and, if so, how far – AI impact assessment
will eliminate the possibility of unfair/illegal discrimination through AI systems. In
any event, discrimination on the basis of protected characteristics requires access to
personal data, which is the topic of the next chapter, on privacy and data protection.

References

Access Now (2018) Human rights in the age of artificial intelligence. Access Now. https://www.
    accessnow.org/cms/assets/uploads/2018/11/AI-and-Human-Rights.pdf. Accessed 1 May 2022
Access Now Policy Team (2018) The Toronto declaration: protecting the right to equality and
    non-discrimination in machine learning systems. Access Now, Toronto. https://www.access
    now.org/cms/assets/uploads/2018/08/The-Toronto-Declaration_ENG_08-2018.pdf. Accessed 26
    Sept 2020
Ahmed M (2020) UK passport photo checker shows bias against dark-skinned women. BBC News,
    8 Oct. https://www.bbc.com/news/technology-54349538. Accessed 4 May 2022
AI HLEG (2019) Ethics guidelines for trustworthy AI. High-level expert group on artificial intel-
    ligence. European Commission, Brussels. https://ec.europa.eu/newsroom/dae/document.cfm?
    doc_id=60419. Accessed 25 Sept 2020
AI HLEG (2020) The assessment list for trustworthy AI (ALTAI). High-level expert group on
    artificial intelligence. European Commission, Brussels. https://ec.europa.eu/newsroom/dae/doc
    ument.cfm?doc_id=68342. Accessed 10 Oct 2020
Becker HA (2001) Social impact assessment. Eur J Oper Res 128:311–321. https://doi.org/10.1016/
    S0377-2217(00)00074-6
Becker HA, Vanclay F (eds) (2003) The international handbook of social impact assessment:
    conceptual and methodological advances. Edward Elgar Publishing, Cheltenham
Brey P, Dainow B (2020) Ethics by design and ethics of use approaches for artificial intelligence,
    robotics and big data. SIENNA. https://ec.europa.eu/info/funding-tenders/opportunities/docs/
    2021-2027/horizon/guidance/ethics-by-design-and-ethics-of-use-approaches-for-artificial-intell
    igence_he_en.pdf
Cambridge Dictionary (n.d.) Discrimination. https://dictionary.cambridge.org/dictionary/english/
    discrimination. Accessed 4 May 2022
Cavoukian A (2017) Global privacy and security, by design: turning the 'privacy vs. security'
    paradigm on its head. Health Technol 7:329–333. https://doi.org/10.1007/s12553-017-0207-1

CEN-CENELEC (2017) Ethics assessment for research and innovation, part 2: ethical impact assessment framework. CWA 17145-2. European Committee for Standardization, Brussels. http://ftp.cencenelec.eu/EN/ResearchInnovation/CWA/CWA17214502.pdf. Accessed 6 Oct 2020

CNIL (2015) Privacy impact assessment (PIA): methodology. Commission Nationale de l'Informatique et des Libertés, Paris

Collins PH, Bilge S (2020) Intersectionality. Wiley, New York

Courtland R (2018) Bias detectives: the researchers striving to make algorithms fair. Nature 558:357–360. https://doi.org/10.1038/d41586-018-05469-3

ECP (2019) Artificial intelligence impact assessment. ECP Platform for the Information Society, The Hague. https://ecp.nl/wp-content/uploads/2019/01/Artificial-Intelligence-Impact-Assessment-English.pdf. Accessed 1 May 2022

EDPS (2020) EDPS opinion on the European Commission's white paper on artificial intelligence: a European approach to excellence and trust (opinion 4/2020). European Data Protection Supervisor, Brussels. https://edps.europa.eu/data-protection/our-work/publications/opinions/edps-opinion-european-commissions-white-paper_en. Accessed 6 May 2022

Equality Act (2010) c15. HMSO, London. https://www.legislation.gov.uk/ukpga/2010/15/contents. Accessed 5 May 2022

European Commission (2021) Proposal for a regulation of the European Parliament and of the council laying down harmonised rules on artificial intelligence (Artificial Intelligence Act) and amending certain union legislative acts. European Commission, Brussels. https://eur-lex.europa.eu/legal-content/EN/TXT/?uri=CELEX%3A52021PC0206. Accessed 1 May 2022

Fothergill BT, Knight W, Stahl BC, Ulnicane I (2019) Intersectional observations of the Human Brain Project's approach to sex and gender. J Inf Commun Ethics Soc 17:128–144. https://doi.org/10.1108/JICES-11-2018-0091

FRA (2020) Getting the future right: artificial intelligence and fundamental rights. European Union Agency for Fundamental Rights, Luxembourg

Friedman B, Kahn P, Borning A (2008) Value sensitive design and information systems. In: Himma K, Tavani H (eds) The handbook of information and computer ethics. Wiley Blackwell, Hoboken, pp 69–102

Gunning D, Stefik M, Choi J et al (2019) XAI: explainable artificial intelligence. Sci Robot 4(37). https://doi.org/10.1126/scirobotics.aay7120

Hartley N, Wood C (2005) Public participation in environmental impact assessment: implementing the Aarhus convention. Environ Impact Assess Rev 25:319–340. https://doi.org/10.1016/j.eiar.2004.12.002

Heilweil R (2019) Artificial intelligence will help determine if you get your next job. Vox-Recode, 12 Dec. https://www.vox.com/recode/2019/12/12/20993665/artificial-intelligence-ai-job-screen. Accessed 4 May

Holzinger A, Biemann C, Pattichis CS, Kell DB (2017) What do we need to build explainable AI systems for the medical domain? arXiv:1712.09923 [cs, stat]. https://doi.org/10.48550/arXiv.1712.09923

ICO (2008) Privacy by design. Information Commissioner's Office, Wilmslow. https://web.archive.org/web/20121222044417if_/http://www.ico.gov.uk:80/upload/documents/pdb_report_html/privacy_by_design_report_v2.pdf. Accessed 6 Oct 2020

IEEE (2020) 7010-2020: IEEE recommended practice for assessing the impact of autonomous and intelligent systems on human well-being. IEEE Standards Association, Piscataway, NJ. https://doi.org/10.1109/IEEESTD.2020.9084219

Ivanova Y (2020) The data protection impact assessment as a tool to enforce non-discriminatory AI. In: Antunes L, Naldi M, Italiano GF et al (eds) Privacy technologies and policy. 8th Annual privacy forum, APF 2020, Lisbon, Portugal, 22–23 Oct. Springer Nature Switzerland, Cham, pp 3–24. https://doi.org/10.1007/978-3-030-55196-4_1

Kaur D (2021) Has artificial intelligence revolutionized recruitment? Tech Wire Asia, 9 Feb. https://techwireasia.com/2021/02/has-artificial-intelligence-revolutionized-recruitment/. Accessed 4 May 2022

Latonero M (2018) Governing artificial intelligence: upholding human rights & dignity. Data & Society. https://datasociety.net/wp-content/uploads/2018/10/DataSociety_Governing_ Artificial_Intelligence_Upholding_Human_Rights.pdf. Accessed 26 Sept 2020

Lentz A (2021) Garbage in, garbage out: is AI discriminatory or simply a mirror of IRL inequalities? 18 Jan. Universal Rights Group, Geneva. https://www.universal-rights.org/blog/garbage-in-gar bage-out-is-ai-discriminatory-or-simply-a-mirror-of-irl-inequalities/. Accessed 4 May 2022

Liberty (n.d.) Predictive policing. https://www.libertyhumanrights.org.uk/fundamental/predictive-policing/. Accessed 4 May 2022

Marx K (2017) Manifest der Kommunistischen Partei. e-artnow

McCarthy OJ (2019) AI & global governance: turning the tide on crime with predictive policing. Centre for Policy Research, United Nations University. https://cpr.unu.edu/publications/articles/ ai-global-governance-turning-the-tide-on-crime-with-predictive-policing.html. Accessed 4 May 2022

Microsoft, Article One (2018) Human rights impact assessment (HRIA) of the human rights risks and opportunities related to artificial intelligence (AI). https://www.articleoneadvisors.com/case-studies-microsoft. Accessed 1 May 2022

Mittelstadt B (2019) Principles alone cannot guarantee ethical AI. Nat Mach Intell 1:501–507. https://doi.org/10.1038/s42256-019-0114-4

Mugari I, Obioha EE (2021) Predictive policing and crime control in the United States of America and Europe: trends in a decade of research and the future of predictive policing. Soc Sci 10:234. https://doi.org/10.3390/socsci10060234

Muller C (2020) The impact of artificial intelligence on human rights, democracy and the rule of law. Ad Hoc Committee on Artificial Intelligence (CAHAI), Council of Europe, Strasbourg. https://rm.coe.int/cahai-2020-06-fin-c-muller-the-impact-of-ai-on-human-rights-democr acy-/16809ed6da. Accessed 2 May 2022

PwC (2019) A practical guide to responsible artificial intelligence. https://www.pwc.com/gx/en/ issues/data-and-analytics/artificial-intelligence/what-is-responsible-ai/responsible-ai-practical-guide.pdf. Accessed 18 June 2020

Reisman D, Schultz J, Crawford K, Whittaker M (2018) Algorithmic impact assessments: a practical framework for public agency accountability. AI Now Institute, New York. https://ainowinstitute. org/aiareport2018.pdf. Accessed 18 June 2020

Reuters (2016) Passport robot tells Asian man his eyes are closed. New York Post, 7 Dec. https://nyp ost.com/2016/12/07/passport-robot-tells-asian-man-his-eyes-are-closed/. Accessed 4 May 2022

Reuters (2018) Amazon ditched AI recruiting tool that favored men for technical job. The Guardian, 11 Oct. https://www.theguardian.com/technology/2018/oct/10/amazon-hiring-ai-gen der-bias-recruiting-engine. Accessed 4 May 2022

Stahl BC (2021) Artificial intelligence for a better future: an ecosystem perspective on the ethics of AI and emerging digital technologies. Springer Nature Switzerland AG, Cham. https://doi.org/ 10.1007/978-3-030-69978-9

UK AI Council (2021) AI roadmap. Office for Artificial Intelligence, London. https://assets.publis hing.service.gov.uk/government/uploads/system/uploads/attachment_data/file/949539/AI_Cou ncil_AI_Roadmap.pdf

UN (1948) Universal declaration of human rights. http://www.un.org/en/universal-declaration-human-rights/. Accessed 4 May 2022

UNESCO (2020) First draft text of the recommendation on the ethics of artificial intelligence, 7 Sept. Ad hoc expert group (AHEG) for the preparation of a draft text, UNESCO, Paris. https:// unesdoc.unesco.org/ark:/48223/pf0000373434. Accessed 12 Oct 2020

van den Hoven J (2013) Value sensitive design and responsible innovation. In: Owen R, Heintz M, Bessant J (eds) Responsible innovation. Wiley, Chichester, pp 75–84

Veale M, Binns R (2017) Fairer machine learning in the real world: mitigating discrimination without collecting sensitive data. Big Data Soc 4(2). https://doi.org/10.1177/2053951717743530

Wexler R (2017a) Code of silence. Washington Monthly, 11 June. https://washingtonmonthly.com/ 2017a/06/11/code-of-silence/. Accessed 4 May 2022

Wexler R (2017b) When a computer program keeps you in jail. The New York Times, 13 June. https://www.nytimes.com/2017b/06/13/opinion/how-computers-are-harming-criminal-jus tice.html. Accessed 4 May 2022

Wright D (2011) A framework for the ethical impact assessment of information technology. Ethics Inf Technol 13:199–226. https://doi.org/10.1007/s10676-010-9242-6

Zheng Y, Walsham G (2021) Inequality of what? An intersectional approach to digital inequality under Covid-19. Inf Organ 31:100341. https://doi.org/10.1016/j.infoandorg.2021.100341

Zou J, Schiebinger L (2018) AI can be sexist and racist: it's time to make it fair. Nature 559:324–326. https://doi.org/10.1038/d41586-018-05707-8

Chapter 3
Privacy

Abstract Privacy and data protection are concerns raised about most digital technologies. The advance of artificial intelligence (AI) has given even higher levels of prominence to these concerns. Three cases are presented as examples to highlight the way in which AI can affect or exacerbate privacy concerns. The first deals with the use of private data in authoritarian regimes. The second looks at the implications of AI use of genetic data. The third concerns problems linked to biometric surveillance. Then follows a description of how privacy concerns are currently addressed via data protection regulation and a discussion of where AI may raise new challenges to existing data protection regimes. Current European data protection law requires data protection impact assessment. This chapter suggests that a broader AI impact assessment could broaden the remit of such an assessment to offer more comprehensive coverage of possible privacy concerns linked to AI.

Keywords Privacy · Data protection · Social credit · Data misuse · Authoritarian government · Genetic data · Biometrics · Surveillance

3.1 Introduction

Concerns about the possible negative impact of artificial intelligence (AI) on privacy have been widely expressed (EDPS 2020). Not all AI applications use personal data and therefore some uses may not have any privacy implications. However, the need for large datasets for the training and validation of machine learning models can raise a range of different concerns. Privacy is a complex concept that we return to in more detail below. Key to the discussion of privacy in AI is the worry that the use of AI technologies can lead to the violation of data protection principles, which then leads to harm for specific individuals or groups whose data is analysed using AI.

Privacy and data protection are issues that apply to most digital technologies, including AI. It is possible for most personal data to be misused for purposes that breach data protection principles or violate legitimate privacy preferences unless appropriate safeguards are in place. An important legal recognition of the "right to privacy" based on legitimate privacy preferences was the first, expressed in the

B. C. Stahl et al., *Ethics of Artificial Intelligence*,
SpringerBriefs in Research and Innovation Governance,
https://doi.org/10.1007/978-3-031-17040-9_3

nineteenth century (Warren and Brandeis 1890). The stipulated "right to be let alone" as described by Warren and Brandeis was driven by a key technical innovation of the time, namely the ability to take photographs of individuals. This new technology raised concerns that had previously been immaterial when capturing the likeness of a person required them to sit down in front of a painter for extended periods.

Ever since the nineteenth century, data protection regulation and legislation have developed in tandem with new technical capabilities and resulting threats to privacy. The growing ability to process data through electronic computers led to much academic debate on the topic and the development of so-called principles of fair information practices (Severson 1997). These were originally developed in the US in 1973. They still underpin much of our thinking on data protection today. The principles include that

1. individuals should have the right to know how organizations use personal information and to inspect their records and correct any errors;
2. individuals should have the right to prevent secondary use of personal information if they object to such use; and
3. organizations that collect or use personal information must take reasonable precautions to prevent misuse of the information. (Culnan 1993: 344)

These principles have contributed to the creation of legislation and shaped its content since the 1970s and 1980s. At the European level, Directive 95/46/EC established a shared approach and visible data protection principles in 1995. It was superseded by the General Data Protection Regulation (GDPR) (European Parliament and Council of the EU 2016), which came into effect in 2018.

Given that AI is not the first potential threat to privacy or data protection, it is worth asking why the impacts of AI technologies on privacy are often seen as key ethical concerns. One part of the answer is that machine learning allows the development of fine-grained categories of data which, in turn, can be used to categorise and profile individuals. Such profiling may well be the intended result of AI use, for instance when an organisation seeks to identify potential customers to target with advertising campaigns. Such profiling may also have discriminatory effects as outlined in Chap. 2. It may also have other undesirable consequences for individuals or groups and open the way to misuse, such as when consumer profiles are used for political purposes (see Chap. 5).

AI uses of personal data can furthermore facilitate surveillance far beyond the capabilities that existed prior to AI. This includes automated surveillance of individuals using their biometric data, for example employing facial recognition, as developed in more detail in the cases below. There may be good reasons for the development and employment of such surveillance, as well as morally desirable outcomes, for instance the prevention of gender-based violence. But AI-based surveillance may also have undesired outcomes. The key challenge is that data protection is a moral value that must be balanced against other moral values. This is important to keep in mind from a moral perspective, especially because data protection is strongly regulated whereas other ethical issues and possible moral advantages are typically not subject to the same level of regulation. The following cases of privacy violations that are enabled by AI demonstrate this point.

3.2 Cases of Privacy Violations Through AI

3.2.1 Case 1: Use of Personal Data by Authoritarian Regimes

China is one of the world's leading nations in AI development. It embraces the use of large amounts of data that it collects on its citizens, for instance in its social credit scoring system. This system uses a large number of data points, including social media data, local government data and citizens' activities, to calculate a trustworthiness score for every citizen. Several data platforms are used to integrate data into "a state surveillance infrastructure" (Liang et al. 2018). High scores lead to the allocation of benefits, such as lower utility rates and favourable booking conditions, whereas low scores can lead to the withdrawal of services (Raso et al. 2018). Within China, the system benefits from high levels of approval because Chinese citizens "interpret it through frames of benefit-generation and promoting honest dealings in society and the economy instead of privacy-violation" (Kostka 2019: 1565).

All states collect information about their citizens for a broad range of purposes. Some of these purposes may enjoy strong support from citizens, such as the allocation of financial support or healthcare, while others may be less popular, such as tax collection. Authoritarian governments can make additional use of data on their citizens to stabilise their power base (Liu 2019). A case in point is China, even though research has shown that Chinese citizens interpret the system from the perspective of its benefits.

It has also been argued that China has strong data protection laws. However, these do not apply to state bodies (Gal 2020), and government use of data for schemes such as social credit scoring are therefore not covered. This differs from the situation in Europe, where data protection law is binding on governments and state bodies as well. Social credit scoring is contentious. However, it is not always too different from activities such as "nudging" that democratic governments use, for example to encourage healthy behaviour such as giving up smoking or taking up exercise (Benartzi et al. 2017).

Both nudging and social credit scoring are contested, though one can see arguments in their favour. But the use of AI for the repression of citizens can go far beyond these. China, for example, has been reported to use AI to track behaviour that is deemed suspicious, such as religious speech by its Uighur population (Andersen 2020). Uighur community members who live in the Xianjiang area are subject to intrusive data collection and analysis that checks not only whether they exchange religious texts, but also where they live, their movement patterns, their pregnancy status and much more. The intention behind this data collection is ostensibly to strengthen the state's control of the Uighur community. China's human rights record, in particular with regard to the Uighurs, suggests that this use of data and AI analysis is likely to result in further reduction of freedoms and limiting of human rights. By employing AI, authoritarian regimes may find it easier to analyse large amounts of data, such as social media posts, and to identify contributions that can trigger government responses.

3.2.2 Case 2: Genetic Privacy

> Many genetic programmes are hailed for delivering medical breakthroughs via personalised medicine and the diagnosis of hereditary diseases. For instance, the Saudi Human Genome Program (SHGP), launched by the Saudi King in 2013, was announced with such aims (Alrefaei et al. 2022). Research showed that "90.7% of [Saudi] participants agreed that AI could be used in the SHGP" (ibid). However, the same research showed "a low level of knowledge … regarding sharing and privacy of genetic data" (ibid), pointing to a potential mismatch of awareness of the benefits as opposed to the risks of genetic research supported by AI.

Genetic data is data that can provide deep insights into medical conditions, but also regarding possible risks and propensities for diseases that can go beyond other types of data. It thus has the properties of medical data and is therefore subject to stronger data protection regimes as part of a special category of data in many jurisdictions. Yet the importance and potential of genetic data goes beyond its medical uses. Genetic data of one person can provide information about their human heritage, their ancestors and their offspring. Access to genetic data can therefore present benefits as well as risks and entail a multitude of ethical issues. For instance, genomic datasets can improve research on cancer and rare diseases, while the reidentification of even anonymised data risks serious privacy concerns for the families involved (Takashima et al. 2018).

With the costs of gene sequencing continuing to fall, one can reasonably expect genetic data to become part of routine healthcare within a decade. This raises questions about data governance, storage, security etc. Such genetic data requires Big Data analytics approaches typically based on some sort of AI in order to be viable and provide relevant scientific or diagnostic insights.

In addition to the use of genetic data in healthcare, there is a growing number of private providers, such as 23andMe, Ancestry and Veritas Genetics (Rosenbaum 2018) that offer gene sequencing services commercially. This raises further questions around the ownership of data and the security of these companies, and creates uncertainty about the use of data should such a company go bankrupt or be bought out.

Addressing ethical concerns can lead to unpleasant surprises, for example when a genetic analysis contradicts assumed relationships in a family, proving that someone's ancestry is not as had been supposed. In some cases this may be greeted with humour or mild embarrassment, but in others, where ancestry is crucial to the legitimacy of a social position, evidence of this kind may have manifestly negative consequences. Such consequences, it could be argued, are part of the nature of genetic data and should be dealt with via appropriate information and consent procedures. However, it is in the nature of genetic data that it pertains to more than one individual. If a sibling, for example, undertakes a genetic analysis, then many of the findings will be relevant to other family members. If such an analysis shows, for instance, that a parent is carrying a gene that contributes to a disease, other siblings' propensities to develop this disease would likely be increased as well, even though they did not take a genetic test themselves. This example demonstrates the possible conflicts arising from possessing and sharing such information.

AI analysis of genetic data may lead to medical insights. Indeed, this is the assumption that supports the business model of private gene-sequencing organisations. Their work is built on the assumption that collecting large amounts of genetic data in addition to other data that their customers provide will allow them to identify genetic patterns that can help predict or explain diseases. This, in turn, opens the way to medical research and finding cures, potentially a highly lucrative business.

From an ethical perspective this is problematic because the beneficiaries of this data analysis will normally be the companies, whereas the individual data subjects or donors will at best be notified of the insights their data has contributed to. Another concern is that the analysis may lead to the ability to predict disease trajectories without being able to intervene or cure (McCusker and Loy 2017), thus forcing patients to face difficult decisions involving complex probabilities that most non-experts are poorly equipped to deal with.

A further concern is that of mission creep, where the original purpose of the data collection is replaced by a changing or altogether different use. One obvious example is the growing interest from law enforcement agencies in gaining access to more genetic data so that they can, for example, identify culprits through genetic fingerprinting. The main point is that data, once it exists in digital form, is difficult to contain. Moor (2000) uses the metaphor of grease in an internal combustion engine. Data, once in an electronically accessible format, is very difficult to remove, just like grease in an engine. It may end up in unexpected places, and attempts to delete it may prove futile. In the case of genetic data this raises problems of possible future, and currently unpredicted, use, which, due to the very personal nature of the data, may have significant consequences that one currently cannot predict.

The Saudi case is predicated on the assumption of beneficial outcomes of the sharing of genetic data, and so far there is little data to demonstrate whether and in what way ethical issues have arisen or are likely to arise. A key concern here is that due to the tendency of data to leak easily, waiting until ethical concerns have materialised before addressing them is unlikely to be good enough. At that point the genie will be out of the bottle and the "greased" data may be impossible to contain.

3.2.3   Case 3: Biometric Surveillance

"Nijeer Parks is the third person known to be arrested for a crime he did not commit based on a bad face recognition match" (Hill 2020). Parks was falsely accused of stealing and trying to hit a policy officer with his car based on facial recognition software – but he was 30 miles away at the time. "Facial recognition ... [is] very good with white men, very poor on Black women and not so great on white women, even" (Balli 2021). It becomes particularly problematic when "the police trust the facial recognition technology more than the individual" (ibid).

Biometric surveillance uses data about the human body to closely observe or follow an individual. The most prominent example of this is the use of facial features in order to track someone. In this broad sense of the term, any direct observation of a person, for example a suspected criminal, is an example of biometric surveillance. The main reason why biometric surveillance is included in the discussion of privacy concerns is that AI systems allow an enormous expansion of the scope of such surveillance. Whereas in the past one observer could only follow one individual, or maybe a few, the advent of machine learning and image recognition techniques, coupled with widespread image capture from closed-circuit television cameras, allows community surveillance. Automatic face recognition and tracking is not the only possible example of biometric surveillance, but it is the one that is probably most advanced and raises most public concern relating to privacy, as in the case described above.

There are numerous reasons why biometric surveillance is deemed to be ethically problematic. It can be done without the awareness of the data subject and thus lead to the possibility and the perception of pervasive surveillance. While some might welcome pervasive surveillance as a contribution to security and the reduction of crime, it has been strongly argued that being subject to it can lead to significant harm. Brown (2000), drawing on Giddens (1984) and others, argues that humans need a "protective cocoon" that shields from external scrutiny. This is needed to develop a sense of "ontological security", a condition for psychological and mental health. Following this argument, pervasive surveillance is ethically problematic, simply for the psychological damage it can do through its very existence. Surveillance can lead to self-censoring and "social cooling" (Schep n.d.), that is, a modification of social interaction caused by fear of possible sanctions. AI-enabled large-scale biometric surveillance could reasonably be expected to lead to this effect.

3.3   Data Protection and Privacy

The above three case studies show that the analysis of personal data through AI systems can lead to significant harms. AI is by far not the only threat to privacy, but it adds new capabilities that can either exacerbate existing threats, for example by automating mass surveillance based on biometric data, or add new angles to privacy

Fig. 3.1 Seven types of privacy

concerns, for example by exposing new types of data, such as genetic data, to the possibility of privacy violations.

Before we look at what is already being done to address these concerns and what else could be done, it is worth providing some more conceptual clarity. The title of this chapter and the headlines covering much of the public debate on the topics raised here refer to "privacy". As suggested at the beginning of this chapter, however, privacy is a broad term that covers more than the specific aspects of AI-enabled analysis of personal data.

A frequently cited categorisation of privacy concepts (Finn et al. 2013: 7) proposes that there are seven types of privacy: privacy of the person, privacy of behaviour and action, privacy of personal communication, privacy of data and image, privacy of thoughts and feelings, privacy of location and space, and privacy of association (including group privacy) (see Fig. 3.1).

Most of these types of privacy have can be linked to data, but they go far beyond simple measures of data protection. Nissenbaum (2004) suggests that privacy can be understood as contextual integrity. This means that privacy protection must be context-specific and that information gathering needs to conform to the norms of the context. She uses this position to argue against public surveillance.

It should thus be clear that privacy issues cannot be comprehensively resolved by relying on formal mechanisms of data protection governance, regulation and/or legislation. However, data protection plays a crucial role in and is a necessary condition of privacy preservation. The application of data protection principles to AI raises several questions. One relates to the balance between the protection of personal data and the openness of data for novel business processes, where it has been argued that stronger data protection rules, such as the EU's GDPR (European Parliament and Council of the EU 2016), can lead to the weakening of market positions in the race for AI dominance (Kaplan and Haenlein 2019). On the other hand, there are worries that current data protection regimes may not be sufficient in their coverage to deal with novel privacy threats arising from AI technologies and applications (Veale et al. 2018).

A core question which has long been discussed in the broader privacy debate is whether privacy is an intrinsic or an instrumental value. Intrinsic values are those values that are important in themselves and need no further justification. Instrumental values are important because they lead to something that is good (Moor 2000). The distinction is best known in environmental philosophy, where some argue that an intact natural environment has an intrinsic value while others argue that it is solely needed for human survival or economic reasons (Piccolo 2017).

However, this distinction may be simplistic, and the evaluation of a value may require attention to both intrinsic and instrumental aspects (Sen 1988). For our purposes it is important to note that the question whether privacy is an intrinsic or instrumental value has a long tradition (Tavani 2000). The question is not widely discussed in the AI ethics discourse, but the answer to it is important in determining the extent to which AI-related privacy risks require attention. The recognition of privacy as a fundamental right, for example in the European Charter of Funda-mental Rights (European Union 2012), settles this debate to some degree and posits privacy as a fundamental right worthy of protection (AI HLEG 2020). However, even assuming that privacy is an unchanging human right, technology will affect how respect for privacy is shown (Buttarelli 2017). AI can also raise novel threats to privacy, for example by making use of emotion data that do not fit existing remedies (Dignum 2019).

Finally, like most other fundamental rights, privacy is not an absolute right. Personal privacy finds its limits when it conflicts with other basic rights or obli-gations, for example when the state compiles data in order to collect taxes or prevent the spread of diseases. The balancing of privacy against other rights and obliga-tions therefore plays an important role in finding appropriate mitigations for privacy threats.

3.4 Responses to AI-Related Privacy Threats

We propose two closely related responses to AI-related privacy threats: data protection impact assessments (DPIAs) and AI impact assessments (AI-IAs).

DPIAs developed from earlier privacy impact assessments (Clarke 2009; ICO 2009). They are predicated on the idea that it is possible to proactively identify possible issues and address them early in the development of a technology or a socio-technical system. This idea is widespread and there are numerous types of impact assessment, such as environmental impact assessments (Hartley and Wood 2005), social impact assessments (Becker and Vanclay 2003) and ethics impact assessments (Wright 2011). The choice of terminology for DPIAs indicates a recognition of the complexity of the concept of privacy and a consequently limited focus on data protection only. DPIAs are mandated in some cases under the GDPR. As a result of this legal requirement, DPIAs have been widely adopted and there are now well-established methods that data controllers can use.

Data Controllers and Data Processors

The concept of a data controller is closely linked to the GDPR, where it is defined as the organisation that determines the purposes for which and the means by which personal data is processed. The data controller has important responsibilities with regard to the data they control and is normally liable when data protection rules are violated. The data processor is the organisation that processes personal data on behalf of the data controller. This means that data controller and data processor have clearly defined tasks, which are normally subject to a contractual agreement. An example might be a company that analyses personal data for training a machine learning system. This company, because it determines the purpose and means of processing, is the controller. It may store the data on a cloud storage system. The organisation running the cloud storage could then serve as data processor (EDPS 2019; EDPB 2020).

In practice, DPIAs are typically implemented in the form of a number of questions that a data controller or data processor has to answer, in order to identify the type of data and the purpose and legal basis of the data processing, and to explore whether the mechanisms in place to protect the data are appropriate to the risk of data breaches (ICO n.d.). The risk-based approach that underlies DPIAs, or at least those undertaken in response to the GDPR, shows that data protection is not a static requirement but must be amenable to the specifics of the context of data processing. This can raise questions when AI is used for data processing, as the exact uses of machine learning models may be difficult to predict, or where possible harms would not target the individual data subject but may occur at a social level, for instance when groups of the population are stigmatised because of characteristics that are manifest in their personal data. An example might be a healthcare system that identifies a correlation between membership of an ethnic group and propensity to a particular disease. Even though this says nothing about causality, it could nevertheless lead to prejudice against members of the ethnic group.

We have therefore suggested that a broader type of impact assessment is more appropriate for AI, one that includes questions of data protection but also looks at other possible ethical issues in a more structured way. Several such AI-IAs have been developed by various institutions. The most prominent was proposed by the EU's High-Level Expert Group in its Assessment List for Trustworthy AI, or ALTAI (AI HLEG 2020). Other examples are the AI Now Institute's algorithmic impact assessment (Reisman et al. 2018), the IEEE's recommended practice for assessing the impact of autonomous and intelligent systems on human wellbeing (IEEE 2020) and the ECP Platform's artificial intelligence impact assessment (ECP 2019).[1]

What all these examples have in common is that they broaden the idea of an impact assessment for AI to address various ethical issues. They all cover data protection questions but go beyond them. This means that they may deal with questions of long-term or large-scale use of AI, such as economic impact or changes in democratic norms, that go beyond the protection of individual personal data. In fact, there are several proposals that explicitly link AI-IAs and DPIAs or that focus in particular on the data protection aspect of an AI-IA (Government Digital Service 2020; Ivanova

---

[1] A collection of recent AI impact assessments can be accessed via a public Zotero group at https://www.zotero.org/groups/4042832/ai_impact_assessments.

2020; ICO 2021). An AI-IA should therefore not be seen as a way to replace a DPIA, but rather as supplementing and strengthening it.

3.5 Key Insights

Privacy remains a key concern in the AI ethics debate and this chapter has demonstrated several ways in which AI can cause harm, based on the violation of data protection principles. Unlike other aspects of the AI ethics debate, privacy is recognised as a human right, and data protection, as a means of supporting privacy, is extensively regulated. As a result of this high level of attention, there are well-established mechanisms, such as DPIAs, which can easily be extended to cover broader AI issues or incorporated into AI-IAs.

The link between DPIAs and AI-IAs can serve as an indication of the role of data and data protection as a foundational aspect of many other ethical issues. Not all of AI ethics can be reduced to data protection. However, many of the other issues discussed in this book have a strong link to personal data. Unfair discrimination, for example, typically requires and relies on personal data on the individuals who are discriminated against. Economic exploitation in surveillance capitalism is based on access to personal data that can be exploited for commercial purposes. Political and other types of manipulation require access to personal data to identify personal preferences and propensities to react to certain stimuli. Data protection is thus a key to many of the ethical issues of AI, and our suggested remedies are therefore likely to be relevant across a range of issues. Many of the responses to AI ethics discussed in this book will, in turn, touch on or incorporate aspects of data protection.

This does not imply, however, that dealing with privacy and data protection in AI is easy or straightforward. The responses that we suggest here, i.e. DPIAs and AI-IAs, are embedded in the European context, in which privacy is recognised as a human right and data protection has been codified in legislation. It might be challenging to address such issues in the absence of this societal and institutional support. Our Case 1 above, which describes the use of data by an authoritarian regime, is a reminder that state and government-level support for privacy and data protection cannot be taken for granted.

Another difficulty lies in the balancing of competing goods and the identification of the boundaries of what is appropriate and ethically defensible. We have mentioned the example of using AI to analyse social media to identify cases of religious speech that can be used to persecute religious minorities. The same technology can be used to search social media in a different institutional context to identify terrorist activities. These two activities may be technically identical, though they are subject to different interpretations. This raises non-trivial questions about who determines what constitutes an ethically legitimate use of AI and where the boundaries of that use are, and on what grounds such distinctions are drawn. This is a reminder that AI ethics can rarely be resolved simply, but needs to be interpreted from a broader perspective that includes a systems view of the AI application and considers institutional and societal aspects when ethical issues are being evaluated and mitigated.

References

AI HLEG (2020) The assessment list for trustworthy AI (ALTAI). High-level expert group on artificial intelligence. European Commission, Brussels. https://ec.europa.eu/newsroom/dae/doc ument.cfm?doc_id=68342. Accessed 10 Oct 2020

Alrefaei AF, Hawsawi YM, Almaleki D et al (2022) Genetic data sharing and artificial intelligence in the era of personalized medicine based on a cross-sectional analysis of the Saudi human genome program. Sci Rep 12:1405. https://doi.org/10.1038/s41598-022-05296-7

Andersen R (2020) The panopticon is already here. The Atlantic, Sept. https://www.theatlantic.com/magazine/archive/2020/09/china-ai-surveillance/614197/. Accessed 7 May 2022

Balli E (2021) The ethical implications of facial recognition technology. ASU News, 17 Nov. https://news.asu.edu/20211117-solutions-ethical-implications-facial-recognition-technology. Accessed 7 May 2022

Becker HA, Vanclay F (eds) (2003) The international handbook of social impact assessment: conceptual and methodological advances. Edward Elgar Publishing, Cheltenham

Benartzi S, Besears J, Mlikman K et al (2017) Governments are trying to nudge us into better behavior. Is it working? The Washington Post, 11 Aug. https://www.washingtonpost.com/news/wonk/wp/2017/08/11/governments-are-trying-to-nudge-us-into-better-behavior-is-it-working/. Accessed 1 May 2022

Brown WS (2000) Ontological security, existential anxiety and workplace privacy. J Bus Ethics 23:61–65. https://doi.org/10.1023/A%3A1006223027879

Buttarelli G (2017) Privacy matters: updating human rights for the digital society. Health Technol 7:325–328. https://doi.org/10.1007/s12553-017-0198-y

Clarke R (2009) Privacy impact assessment: its origins and development. Comput Law Secur Rev 25:123–135. https://doi.org/10.1016/j.clsr.2009.02.002

Culnan M (1993) "How did they get my name?" An exploratory investigation of consumer attitudes toward secondary information use. MIS Q 17(3):341–363. https://doi.org/10.2307/249775

Dignum V (2019) Responsible artificial intelligence: how to develop and use AI in a responsible way. Springer Nature Switzerland AG, Cham

ECP (2019) Artificial intelligence impact assessment. ECP Platform for the Information Society, The Hague. https://ecp.nl/wp-content/uploads/2019/01/Artificial-Intelligence-Impact-Assessment-English.pdf. Accessed 1 May 2022

EDPB (2020) Guidelines 07/2020 on the concepts of controller and processor in the GDPR. European Data Protection Board, Brussels. https://edpb.europa.eu/sites/default/files/consultation/edpb_guidelines_202007_controllerprocessor_en.pdf. Accessed 8 May 2022

EDPS (2019) EDPS guidelines on the concepts of controller, processor and joint controllership under regulation (EU) 2018/1725. European Data Protection Supervisor, Brussels

EDPS (2020) EDPS opinion on the European Commission's white paper on artificial intelligence: a European approach to excellence and trust (Opinion 4/2020). European Data Protection Supervisor, Brussels. https://edps.europa.eu/data-protection/our-work/publications/opinions/edps-opinion-european-commissions-white-paper_en. Accessed 6 May 2022

European Parliament, Council of the EU (2016) Regulation (EU) 2016/679 of the European Parliament and of the council of 27 April 2016 on the protection of natural persons with regard to the processing of personal data and on the free movement of such data, and repealing Directive 95/46/EC (General Data Protection Regulation). Official J Eur Union L119(11):1–88. https://eur-lex.europa.eu/legal-content/EN/TXT/PDF/?uri=CELEX:32016R0679&from=EN. Accessed 1 May 2022.

European Union (2012) Charter of fundamental rights of the European Union. https://eur-lex.europa.
   eu/legal-content/EN/TXT/PDF/?uri=CELEX:C2012/326/02&from=EN. Accessed 1 May 2022
Finn RL, Wright D, Friedewald M (2013) Seven types of privacy. In: Gutwirth S, Leenes R, de Hert
   P, Poullet Y (eds) European data protection: coming of age. Springer, Dordrecht, pp 3–32
Gal D (2020) China's approach to AI ethics. In: Elliott H (ed) The AI powered state: China's
   approach to public sector innovation. Nesta, London, pp 53–62
Giddens A (1984) The constitution of society: outline of the theory of structuration. Polity,
   Cambridge
Government Digital Service (2020) Data ethics framework. Central Digital and Data
   Office, London. https://assets.publishing.service.gov.uk/government/uploads/system/uploads/att
   achment_data/file/923108/Data_Ethics_Framework_2020.pdf. Accessed 1 May 2022
Hartley N, Wood C (2005) Public participation in environmental impact assessment: implementing
   the Aarhus convention. Environ Impact Assess Rev 25:319–340. https://doi.org/10.1016/j.eiar.
   2004.12.002
Hill K (2020) Another arrest, and jail time, due to a bad facial recognition match. The New
   York Times, 29 Dec. https://www.nytimes.com/2020/12/29/technology/facial-recognition-mis
   identify-jail.html. Accessed 7 May 2022
ICO (2009) Privacy impact assessment handbook, v. 2.0. Information Commissioner's
   Office, Wilmslow. https://www.huntonprivacyblog.com/wp-content/uploads/sites/28/2013/09/
   PIAhandbookV2.pdf. Accessed 6 Oct 2020
ICO (2021) AI and data protection risk toolkit beta. Information Commissioner's Office, Wilmslow
ICO (n.d.) Data protection impact assessments. Information Commissioner's Office, Wilm-
   slow. https://ico.org.uk/for-organisations/guide-to-data-protection/guide-to-the-general-data-pro
   tection-regulation-gdpr/accountability-and-governance/data-protection-impact-assessments/.
   Accessed 8 May 2022
IEEE (2020) 7010-2020: IEEE recommended practice for assessing the impact of autonomous and
   intelligent systems on human well-being. IEEE Standards Association, Piscataway. https://doi.
   org/10.1109/IEEESTD.2020.9084219
Ivanova Y (2020) The data protection impact assessment as a tool to enforce non-discriminatory
   AI. In: Antunes L, Naldi M, Italiano GF et al (eds) Privacy technologies and policy. 8th Annual
   privacy forum, APF 2020, Lisbon, Portugal, 22–23 Oct. Springer Nature Switzerland, Cham, pp
   3–24. https://doi.org/10.1007/978-3-030-55196-4_1
Kaplan A, Haenlein M (2019) Siri, Siri, in my hand: who's the fairest in the land? On the interpretations, illustrations, and implications of artificial intelligence. Bus Horiz 62:15–25. https://doi.
   org/10.1016/j.bushor.2018.08.004
Kostka G (2019) China's social credit systems and public opinion: explaining high levels of approval.
   New Media Soc 21:1565–1593. https://doi.org/10.1177/1461444819826402
Liang F, Das V, Kostyuk N, Hussain MM (2018) Constructing a data-driven society: China's social
   credit system as a state surveillance infrastructure. Policy Internet 10:415–453. https://doi.org/
   10.1002/poi3.183
Liu C (2019) Multiple social credit systems in China. Social Science Research Network, Rochester
McCusker EA, Loy CT (2017) Huntington disease: the complexities of making and disclosing a
   clinical diagnosis after premanifest genetic testing. Tremor Other Hyperkinet Mov (NY) 7:467.
   https://doi.org/10.7916/D8PK0TDD
Moor JH (2000) Toward a theory of privacy in the information age. In: Baird RM, Ramsower RM,
   Rosenbaum SE (eds) Cyberethics: social and moral issues in the computer age. Prometheus,
   Amherst, pp 200–212
Nissenbaum H (2004) Symposium: privacy as contextual integrity. Wash Law Rev 79:119–158.
   https://digitalcommons.law.uw.edu/cgi/viewcontent.cgi?article=4450&context=wlr. Accessed 2
   May 2022
Piccolo JJ (2017) Intrinsic values in nature: objective good or simply half of an unhelpful dichotomy?
   J Nat Conserv 37:8–11. https://doi.org/10.1016/j.jnc.2017.02.007

Raso FA, Hilligoss H, Krishnamurthy V et al (2018) Artificial intelligence & human rights: opportunities & risks. Berkman Klein Center Research Publication No. 2018-6. https://doi.org/10.2139/ssrn.3259344

Reisman D, Schultz J, Crawford K, Whittaker M (2018) Algorithmic impact assessments: a practical framework for public agency accountability. AI Now Institute, New York. https://ainowinstitute.org/aiareport2018.pdf. Accessed 18 June 2020

Rosenbaum E (2018) 5 biggest risks of sharing your DNA with consumer genetic-testing companies. CNBC, 16 June. https://www.cnbc.com/2018/06/16/5-biggest-risks-of-sharing-dna-with-consumer-genetic-testing-companies.html. Accessed 7 May 2022

Schep T (n.d.) Social cooling. https://www.tijmenschep.com/socialcooling/. Accessed 8 May 2022

Sen A (1988) On ethics and economics, 1st edn. Wiley-Blackwell, Oxford

Severson RW (1997) The principles for information ethics, 1st edn. Routledge, Armonk

Takashima K, Maru Y, Mori S et al (2018) Ethical concerns on sharing genomic data including patients' family members. BMC Med Ethics 19:61. https://doi.org/10.1186/s12910-018-0310-5

Tavani H (2000) Privacy and security. In: Langford D (ed) Internet ethics, 2000th edn. Palgrave, Basingstoke, pp 65–95

Veale M, Binns R, Edwards L (2018) Algorithms that remember: model inversion attacks and data protection law. Phil Trans R Soc A 376:20180083. https://doi.org/10.1098/rsta.2018.0083

Warren SD, Brandeis LD (1890) The right to privacy. Harv Law Rev 4(5):193–220. https://doi.org/10.2307/1321160

Wright D (2011) A framework for the ethical impact assessment of information technology. Ethics Inf Technol 13:199–226. https://doi.org/10.1007/s10676-010-9242-6

Chapter 4
Surveillance Capitalism

Abstract Surveillance capitalism hinges on the appropriation and commercialisation of personal data for profit-making. This chapter spotlights three cases connected to surveillance capitalism: data appropriation, monetisation of health data and the unfair commercial practice when "free" isn't "free". It discusses related ethical concerns of power inequality, privacy and data protection, and lack of transparency and explainability. The chapter identifies responses to address concerns about surveillance capitalism and discusses three key responses put forward in policy and academic literature and advocated for their impact and implementation potential in the current socio-economic system: antitrust regulation, data sharing and access, and strengthening of data ownership claims of consumers/individuals. A combination of active, working governance measures is required to stem the growth and ill-effects of surveillance capitalism and protect democracy.

Keywords Antitrust · Big Tech · Data appropriation · Data monetisation · Surveillance capitalism · Unfair commercial practices

4.1 Introduction

The flourishing of systems and products powered by artificial intelligence (AI) and increasing reliance on them fuels surveillance capitalism and "ruling by data" (Pistor 2020). The main beneficiaries, it is argued, are 'Big Tech' – including Apple, Amazon, Alphabet, Meta, Netflix, Tesla (Jolly 2021) – who are seen to be entrenching their power over individuals and society and affecting democracy (ibid).

Simply described, surveillance capitalism hinges on the appropriation and commercialisation of personal data for profit. Zuboff (2015) conceptualised and defined it as a "new form of information capitalism [which] aims to predict and modify human behavior as a means to produce revenue and market control". She argues that surveillance capitalism "effectively exile[s] persons from their own behavior while producing new markets of behavioral prediction and modification" (ibid). It is underpinned by organisational use of behavioural data leading to asymmetries in knowledge and power (Zuboff 2019b). As a result, consumers often may not realise the extent to which they are responding to prompts driven by commercial interests.

Doctorow (2021) elaborates that surveillance capitalists engage in segmenting (targeting based on behaviour/attitudes/choices), deception (by making fraudulent claims, replacing beliefs with false or inaccurate ones) and domination (e.g. Google's dominance on internet searches and the monopolisation of the market through mergers and acquisitions). He outlines three reasons why organisations continue to over-collect and over-retain personal data: first, that they are competing with people's ability to resist persuasion techniques once they are wise to them, and with competitors' abilities to target their customers; second, that the cheapness of data aggregation and storage facilitates the organisation's acquiring an asset for future sales; and third, that the penalties are imposed for leaking data are negligible (Wolff and Atallah 2021).

Surveillance capitalism manifests in both the private and public sectors; organisations in both sectors collect and create vast reserves of personal data under various guises. Examples for the private sector can be found in online marketplaces (e.g. Amazon—Nash 2021), the social media industry (e.g. Facebook—Zuboff 2019a) and the entertainment industry (Bellanova and González Fuster 2018). The most cited examples of surveillance capitalism are Google's (Zuboff 2020) and Facebook's (Zuboff 2019a) attempts to feed their systems and services. In the public sector, this is very notable in the healthcare and retail sectors. During the COVID-19 pandemic, attention has been drawn to health-related surveillance capitalism. For example, it has been argued that telehealth technologies were pushed too quickly during the COVID-19 pandemic (Cosgrove et al. 2020; Garrett 2021).

4.2 Cases of AI-Enabled Surveillance Capitalism

4.2.1 Case 1: Data Appropriation

Clearview AI is a software company headquartered in New York which specialises in facial recognition software, including for law enforcement. The company, which holds ten billion facial images, aims to obtain a further 90 billion, which would amount to 14 photos of each person on the planet (Harwell 2022). In May 2021, legal complaints were filed against Clearview AI in France, Austria, Greece, Italy and the United Kingdom. It was argued that photos were harvested from services such as Instagram, LinkedIn and YouTube in contravention of what users of these services were likely to expect or have agreed to (Campbell 2021). On 16 December 2021, the French Commission Nationale de l'Informatique et des Libertés announced that it had "ordered the company to cease this illegal processing and to delete the data within two months" (CNIL 2021).

The investigations carried out by the Commission Nationale de l'Informatique et des Libertés (CNIL) into Clearview AI revealed two breaches of the General Data Protection Regulation (GDPR) (European Parliament and Council of the EU 2016): first, an unlawful processing of personal data (ibid: Article 6), as the collection and use of biometric data had been carried out without a legal basis; and second, a failure to take into account the rights of individuals in an effective and satisfactory way, especially with regard to access to their data (ibid: Articles 12, 15 and 17).

CNIL ordered Clearview AI to stop the collection and use of data of persons on French territory in the absence of a legal basis, to facilitate the exercise of individuals' rights and to comply with requests for erasure. The company was given two months to comply with the injunctions and justify its compliance or face sanctions.[1]

This case is an example of data appropriation (i.e. the illegal, unauthorised or unfair taking or collecting of personal data for known or unknown purposes, without consent or with coerced or uninformed consent). Organisations appropriating data in such a fashion do not offer data subjects "comparable compensation", while the organisations themselves gain commercial profits and other benefits from the activity.

4.2.2 Case 2: Monetisation of Health Data

In 2021, the personal data of 61 million people became publicly available without password protection, due to data leaks at a New York-based provider of health tracking services. The data included personal information such as names, gender, geographic locations, dates of birth, weight and height (Scroxton 2021). Security researcher Jeremiah Fowler, who discovered the database, traced its origin to a company that offered devices and apps to track health and wellbeing data. The service users whose personal data had been leaked were located all over the world. Fowler contacted the company, which thanked him and confirmed that the data had now been secured (Fowler n.d.).

This case highlights issues with the collection and storage of health data on a vast scale by companies using health and fitness tracing devices, and it reveals the vulnerability of such data to threats and exposure.

A related concern is the procurement of such data by companies such as Google through their acquisition of businesses such as Fitbit, a producer of fitness monitors and related software. Experts (Bourreau et al. 2020) have indicated that such acquisition is problematic for various reasons: major risks of "platform envelopment", the extension of monopoly power (by undermining competition) and consumer exploitation. Their concerns also relate to the serious harms that might result from Google's ability to combine its own data with Fitbit's health data.

---

[1] As of May 2022, the CNIL had not published any update regarding the company's compliance or otherwise.

The European Commission carried out an in-depth investigation (European Commission 2020a) into the acquisition of Fitbit by Google. The concerns that emerged related to advertising, in that the acquisition would increase the already extensive amount of data that Google was able to use for the personalisation of ads, and the resulting difficulty for rivals setting out to match Google's services in the markets for online search advertising. It was argued that the acquisition would raise barriers to entry and expansion by Google's competitors, to the detriment of advertisers, who would ultimately face higher prices and have less choice. The European Commission approved the acquisition of Fitbit by Google under the EU Merger Regulation, conditional on full compliance with a ten-year commitments package offered by Google (European Commission 2020b).

4.2.3   Case 3: Unfair Commercial Practices

In 2021, Italy's Consiglio di Stato (Council of State) agreed with the Autorità Garante della Concorrenza e del Mercato (Italian Competition Authority) and the Tribunale Amministrativo Regionale (Regional Administrative Tribunal) of the Lazio region to sanction Facebook for an unfair commercial practice. Facebook was fined seven million euros for misleading its users by not explaining to them in a timely and adequate manner, during the activation of their accounts, that data would be collected with commercial intent (AGCM 2021).

This case spotlights how companies deceive users into believing they are getting a social media service free of charge when this is not true at all (Coraggio 2021). The problem is aggravated by the companies not communicating well enough to users that their data is the quid pro quo for the use of the service and that the service is only available to them conditional on their making their data available and accepting the terms of service. What is also fuzzily communicated is how and to what extent companies put such data into further commercial use and take advantage of it for targeted advertising purposes. Commentators have gone so far as to state that the people using such services have themselves become the product (Oremus 2018).

4.3   Ethical Questions About Surveillance Capitalism

One of the primary ethical concerns that arise in the context of all three case studies is that of *power inequality*. The power of Big Tech is significant; it has even been compared to that of nation states (Andal 2016; Apostolicas 2019) and is further strengthened by the development and/or acquisition of AI solutions. All three of the cases examined have served to further and enhance the control that AI owners hold. The concentrated power that rests with a handful of big tech companies and the control and influence they have, for example on political decision-making (Dubhashi and Lappin 2021), market manipulation and digital lives, are disrupting economic processes (Fernandez et al 2020) and posing a threat to democracy (Fernandez 2021), freedoms of individuals and political and social life.

Another key ethical concern brought to the forefront with these cases is *privacy and data protection* (see Chap. 3). Privacy is critical to human autonomy and well-being and helps individuals protect themselves from interference in their lives (Nass et al 2009). For instance, leaked personal health data might be appropriated by employers or health insurers and used against the interests of the person concerned. Data protection requires that data be processed in a lawful, fair and transparent manner in addition to being purpose-limited, accurate and retained for a limited time. It also requires that such processing respect integrity, confidentiality and accountability principles.

Lack of transparency and explainability links to data appropriation, data monetisation and unfair commercial practices. While it may seem obvious from a data protection and societal point of view that transparency requirements should be followed by companies that acquire personal data, this imperative faces challenges. Transparency challenges are a result of the structure and operations of the data industry. Transparency is also challenged by the appropriation of transparency values in public relations efforts by data brokers (e.g. Acxiom, Experian and ChoicePoint) to water down government regulation (Crain 2016). Crain (2016) also highlights that transparency may only create an "illusion of reform" and not address basic power imbalances.

Other ethical concerns relate to *proportionality and do-no-harm*. The UNESCO Recommendation on the Ethics of Artificial Intelligence, as adopted (UNESCO 2021), suggests:

- The AI method chosen should be appropriate and proportional to achieve a given legitimate aim.
- The AI method chosen should not infringe upon foundational values; its use must not violate or abuse human rights.
- The AI method should be appropriate to the context and should be based on rigorous scientific foundations.

In the cases examined above, there are clear-cut failures to meet these checks. "Appropriateness" refers to whether the technological or AI solution used is the best (with regard to cost and quality justifying any invasions of privacy), whether there is a risk of human rights, such as that to privacy, being abused and data being reused, and whether the objectives can be satisfied using other means. The desirability of the use of AI solutions is also something that should be duly considered– with regard to the purpose, advantages and burden imposed by them on social values, justice and the public interest.

A key aspect of the perception of surveillance capitalism as being ethically problematic seems to be the encroaching of market mechanisms on areas of life that previously were not subject to financial exchange. To some degree this is linked to the perception of exploitation of the data producers. Many users of "free" online services are content to use services such as social media or online productivity tools in exchange for the use of their data by application providers. There is also, nevertheless, a perception of unfairness, as the service providers have been able to make gigantic financial gains that are not shared with the individuals on whose data they rely to generate these gains. In addition, criticism of surveillance capitalism seems to be based on the perception that some parts of social life should be free of market exchange. A manifestation of this may be the use of the term "friend" in social media, where not only does the nature of friendship differ substantially from that in the offline world, but the number of friends and followers can lead to financial transactions that would be deemed inappropriate in the offline world.

There is no single clearly identifiable ethical issue that is at the base of surveillance capitalism. The term should be understood as signifying opposition to technically enabled social changes that concentrate economic and political power in the hands of a few high-profile organisations.

4.4 Responses to Surveillance Capitalism

Many types of responses have been put forward to address concerns about surveillance capitalism: legal or policy-based responses, market-based responses, and societal responses. Legal and policy measures include antitrust regulation, intergovernmental regulation, strengthening the data-ownership claims of consumers or individuals, socialising the ownership of evolving technologies (Garrett 2021), making big tech companies spend their monopoly profits on governance (Doctorow 2021), mandatory disclosure frameworks (Andrew et al. 2021) and greater data sharing and access.

Market-based responses include placing value on the information provided to surveillance capitalists, monopoly reductions (Doctorow 2021), defunding Big Tech and refunding community-oriented services (Barendregt et al. 2021) and users employing their market power by rejecting and avoiding companies with perceived unethical behaviour (Jensen 2019).

Societal responses include indignation (Lyon 2019), naming /public indignation (Kavenna 2019), personal data spaces or emerging intermediary services that allow users control over the sharing and use of their data (Lehtiniemi 2017), increasing data literacy and awareness of how transparent a company's data policy is, and improving consumer education (Lin 2018).

This section examines three responses which present promising ways to curtail the impact of surveillance capitalism and the ethical questions studied in a variety of ways, though none on its own is a silver bullet. These responses have been discussed in policy and academic literature and advocated for their impact and implementation potential in the current socio-economic system. The challenges of surveillance capitalism arise from the socio-political environment in which AI is developed and used, and the responses we have spotlighted here are informed by this.

4.4.1 Antitrust Regulation

"Antitrust" refers to actions to control monopolies, prevent companies from working together to unfairly control prices, and enhance fair business competition. Antitrust laws regulate monopolistic behaviour and prevent unlawful business practices and mergers. Courts review mergers case by case for illegality. Many calls and proposals have been made to counter the power of Big Tech (e.g. Warren 2019). Discussions have proliferated on the use of antitrust regulations to break up big tech companies (The Economist 2019; Waters 2019) and the appointment of regulators to reverse illegal and anti-competitive tech mergers (Rodrigues et al 2020).

Big Tech is regarded as problematic for its concentration of power and control over the economy, society and democracy to the detriment of competition and innovation in small business (Warren 2019). Grunes and Stucke (2015) emphasise the need for competition and antitrust's "integral role to ensure that we capture the benefits of a data-driven economy while mitigating its associated risks". However, the use of antitrust remedies to control dominant firms presents some problems, such as reducing competitive incentives (by forcing the sharing of information) and innovation, creating privacy concerns or resulting in stagnation and fear among platform providers (Sokol and Comerford 2016).

There are both upsides and downsides to the use of antitrust regulation as a measure to curb the power of Big Tech (Zuboff 2021). The upsides include delaying or frustrating acquisitions, generating better visibility, transparency and oversight, pushing Big Tech to improve their practices, and better prospects for small businesses (Warren 2019). One downside is that despite the antitrust actions taken thus far, some Big Tech companies continue to grow their power and dominance (Swartz 2021). Another downside is the implementation and enforcement burdens antitrust places on regulators (Rodrigues et al. 2020). Furthermore, antitrust actions are expensive and disruptive to business and might affect innovation (The Economist 2019).

Developments (legislative proposals, acquisition challenges, lawsuits and fines) in the USA and Europe show that Big Tech's power is under deeper scrutiny than ever before (Reuters 2021; Council of the EU 2021).

4.4.2 Data Sharing and Access

Another response to surveillance capitalism concerns is greater data sharing and access (subject to legal safeguards and restrictions). Making data open and freely available under a strict regulatory environment is suggested as having the potential to better address the limitations of antitrust legislation (Leblond 2020). In similar vein, Kang (2020) suggests that data-sharing mandates (securely enforced through privacy-enhancing technologies) "have become an essential prerequisite for competition and innovation to thrive"; to counter the "monopolistic power derived from data, Big Tech should share what they know – and make this information widely usable for current and potential competitors" (ibid).

At the European Union level, the proposal for the Data Governance Act (European Commission 2020d) is seen as a "first building block for establishing a solid and fair data-driven economy" and "setting up the right conditions for trustful data sharing in line with our European values and fundamental rights" (European Commission 2021).

The Data Governance Act aims to foster the availability of data for more widespread use by increasing trust in data intermediaries and by strengthening data-sharing mechanisms across the EU. It specifies conditions for the reuse, within the European Union, of certain categories of data held by public sector bodies; a notification and supervisory framework for the provision of data sharing services; and a framework for the voluntary registration of entities which collect and process data made available for altruistic purposes.

The European Commission will also set out "a second major legislative initiative, the Data Act, to maximise the value of data for the economy and society" and "to foster data sharing among businesses, and between businesses and governments" (ibid). The proposed Digital Markets Act aims to lay down harmonised rules ensuring contestable and fair markets in the digital sector across the European Union. Gatekeepers will be present, and it is expected that business access to certain data will go through gatekeepers (European Commission 2020c).

4.4.3   Strengthening of Data Ownership Claims of Consumers/Individuals

Another response to surveillance capitalism is to strengthen the data ownership claims of consumers and individuals. Jurcys et al. (2021) argue that even if user-held data is intangible, it meets all the requirements of an "asset" in property laws and that such "data is specifically defined, has independent economic value to the individual, and can be freely alienated" (ibid).

> [T]he economic benefits property law type of entitlements over user-held data are superior over the set of data rights that are afforded by public law instruments (such as the GDPR or the CCPA) to individuals vis-a-vis third-party service providers who hold and benefit enormously from the information about individuals. (ibid)

Fadler and Legner (2021) also suggest that data ownership remains a key concept to clarify rights and responsibilities but should be revisited in the Big Data and analytics context. They identify three distinct types of data ownership – *data*, *data platform* and *data product ownership* – which may guide the definition of governance mechanisms and serve as the basis for more comprehensive data governance roles and frameworks.

As Hummel, Braun and Dabrock outline (2021), the commonality in calls for data ownership relates to modes of controlling how data is used and the ability to channel, constrain and facilitate the flow of data. They also suggest that with regard to the marketisation and commodification of data, ownership has turned out be a double-edged sword, and that using this concept requires reflection on how data subjects can protect their data and share appropriately. They furthermore outline that "even if legal frameworks preclude genuine ownership in data, there remains room to debate whether they can and should accommodate such forms of quasi-ownership" (Hummel et al. 2021).

Challenges that affect this response include the ambiguousness of the concept of ownership, the complexity of the data value cycle and the involvement of multiple stakeholders, as well as difficulty in determining who could or would be entitled to claim ownership in data (Van Asbroeck et al. 2019).

4.5 Key Insights

A combination of active, working governance measures is required to stem the growth and ill effects of surveillance capitalism and protect democracy. As we move forward, there are some key points that should be considered.

Breaking Up with Antitrust Regulation is Hard to Do

While breaking up Big Tech using antitrust regulation might seem like a very attractive proposition, it is challenging and complex (Matsakis 2019). Moss (2019) assesses the potential consequences of breakup proposals and highlights the following three issues:

- Size thresholds could lead to broad restructuring and regulation.
- Breakup proposals do not appear to consider the broader dynamics created by prohibition on ownership of a platform and affiliated businesses.
- New regulatory regimes for platform utilities will require significant thought.

A report from the EU-funded SHERPA project (Rodrigues et al. 2020) also highlights the implementation burdens imposed on legislators (who must define the letter and scope of the law) and on enforcement authorities (who must select appropriate targets for enforcement action and make enforcement decisions.)

Further challenges include the limitations in antitrust enforcement officials' knowledge and the potential impact of ill-advised investigations and prosecutions on markets (Cass 2013), never-ending processes, defining what conduct contravenes antitrust law (ibid), business and growth disruption, and high costs (The Economist 2019; Waters 2019), including the impact on innovation (Sokol and Comerford 2016).

Are "Big Tech's Employees One of the Biggest Checks on Its Power"?

Among the most significant actions that have changed the way digital companies behave and operate has been action taken by employees of such organisations to hold their employers to account over ethical concerns and illegal practices, while in the process risking career, reputation, credibility and even life. Ghaffary (2021) points out that tech employees are uniquely positioned (with their "behind the scenes" understanding of algorithms and company policies) to provide checks and enable the

Fig. 4.1 Unveiled by whistleblowers

scrutiny needed to influence Big Tech. In the AI context, given issues of lack of transparency, this is significant for its potential to penetrate corporate veils.

A variety of issues have been brought to light by tech whistleblowers: misuse or illegal use of data (Cambridge Analytica) (Perrigo 2019), institutional racism, research suppression (Simonite 2021), suppression of the right to organise (Clayton 2021), the falsification of data, a lack of safety controls and the endangerment of life through hosting hate speech and illegal activity (Milmo 2021) (see Fig. 4.1).

Whistleblowing is now seen in the digital and AI context as a positive corporate governance tool (Brand 2020). Laws have been and are being amended to increase whistleblower protections, for instance in New York and the European Union. Reporting by whistleblowers to enforcement bodies is expected to increase as regulators improve enforcement and oversight over AI. This might provide a necessary check on Big Tech. However, whistleblowing comes with its own price, especially for the people brave enough to take this step, and by itself is not enough, given the resources of Big Tech and the high human and financial costs to the individuals who are forced to undertake such activity (Bridle 2018).

Surveillance capitalism may be here to stay, at least for a while, and its effects might be strong and hard in the short to medium term (and longer if not addressed), but as shown above, there are a plethora of mechanisms and tools to address it. In addition to the responses discussed in this chapter, it is important that other measures be duly reviewed for their potential to support ethical AI and used as required – be they market-based, policy- or law-based or societal interventions. Even more important is the need to educate and inform the public about the implications for and adverse effects on society of surveillance capitalism. This is a role that civil society organisations and the media are well placed to support.

References

AGCM (2021) IP330—Sanzione a Facebook per 7 milioni. Autorità Garante della Concorrenza e del Mercato, Rome. Press release, 17 Feb. https://www.agcm.it/media/comunicati-stampa/2021/2/IP330-. Accessed 10 May 2022

Andal S (2016) Tech giants' powers rival those of nation states. The Interpreter, 6 Apr. The Lowly Institute. https://www.lowyinstitute.org/the-interpreter/tech-giants-powers-rival-those-nation-states. Accessed 15 May 2022

Andrew J, Baker M, Huang C (2021) Data breaches in the age of surveillance capitalism: do disclosures have a new role to play? Crit Perspect Account. https://doi.org/10.1016/j.cpa.2021.102396102396

Apostolicas P (2019) Silicon states: how tech titans are acquiring state-like powers. Harv Int Rev 40(4):18–21. https://www.jstor.org/stable/26917261. Accessed 3 May 2022

Barendregt W, Becker C, Cheon E et al (2021) Defund Big Tech, refund community. Tech Otherwise, 5 Feb. https://techotherwise.pubpub.org/pub/dakcci1r/release/1. Accessed 15 May 2022

Bellanova R, González Fuster G (2018) No (big) data, no fiction? Thinking surveillance with/against Netflix. In: Saetnan AR, Schneider I, Green N (eds) The politics and policies of Big Data: Big Data Big Brother? Routledge, London (forthcoming). https://ssrn.com/abstract=3120038. Accessed 3 May 2022

Bourreau M, Caffarra C, Chen Z et al (2020) Google/Fitbit will monetise health data and harm consumers CEPR Policy Insight No. 107, 30 Sept. https://voxeu.org/article/googlefitbit-will-monetise-health-data-and-harm-consumers. Accessed 3 May 2022

Brand V (2020) Corporate whistleblowing, smart regulation and regtech: the coming of the whistlebot? Univ NSW Law J 43(3):801–826. https://doi.org/10.2139/ssrn.3698446

Bridle J (2018) Whistleblowers are a terrible answer to the problems of Big Tech. Wired, 11 June. https://www.wired.co.uk/article/silicon-valley-whistleblowers-james-bridle-book-new-dark-age. Accessed 3 May 2022

Campbell IC (2021) Clearview AI hit with sweeping legal complaints over controversial face scraping in Europe. The Verge, 27 May. https://www.theverge.com/2021/5/27/22455446/clearview-ai-legal-privacy-complaint-privacy-international-facial-recognition-eu. Accessed 10 May 2022

Cass RA (2013) Antitrust for high-tech and low: regulation, innovation, and risk. J Law Econ Policy 9(2): 169–200. https://ssrn.com/abstract=2138254. Accessed 3 May 2022

Clayton J (2021) Silenced no more: a new era of tech whistleblowing? BBC News, 11 Oct. https://www.bbc.com/news/technology-58850064. Accessed 15 May 2022

CNIL (2021) Facial recognition: the CNIL orders Clearview AI to stop reusing photographs available on the Internet. Commission Nationale de l'Informatique et des Libertés, Paris. https://www.cnil.fr/en/facial-recognition-cnil-orders-clearview-ai-stop-reusing-photographs-available-internet. Accessed 10 May 2022

Coraggio G (2021) Facebook is NOT free and users shall be made aware of paying with their personal data. GamingTechLaw, 12 Apr. https://www.gamingtechlaw.com/2021/04/facebook-not-free-personal-data-italian-court.html. Accessed 3 May 2022

Cosgrove L, Karter JM, Morrill Z, McGinley, M (2020) Psychology and surveillance capitalism: the risk of pushing mental health apps during the COVID-19 pandemic. J Humanistic Psychol 60(5):611–625. https://doi.org/10.1177/0022167820937498. Accessed 3 May 2022

Council of the EU (2021) Regulating "Big Tech": council agrees on enhancing competition in the digital sphere. Press release, 25 Nov. https://www.consilium.europa.eu/en/press/press-releases/2021/11/25/regulating-big-tech-council-agrees-on-enhancing-competition-in-the-digital-sphere/. Accessed 3 May 2022

Crain M (2016) The limits of transparency: data brokers and commodification. New Media Soc 20(1):88–104. https://doi.org/10.1177/1461444816657096

Doctorow C (2021) How to destroy surveillance capitalism. Medium Editions. https://onezero.medium.com/how-to-destroy-surveillance-capitalism-8135e6744d59. Accessed 3 May 2022

Dubhashi D, Lappin S (2021) Scared about the threat of AI? It's the big tech giants that need reining in. The Guardian, 16 Dec. https://www.theguardian.com/commentisfree/2021/dec/16/scared-about-the-threat-of-ai-its-the-big-tech-giants-that-need-reining-in. Accessed 15 May 2022

European Commission (2020a) Case M.9660—Google/Fitbit: public version. DG Competition, European Commission, Brussels. https://ec.europa.eu/competition/mergers/cases1/202120/m9660_3314_3.pdf. Accessed 10 May 2022

European Commission (2020b) Mergers: commission clears acquisition of Fitbit by Google, subject to conditions. Press release, 17 Dec. European Commission, Brussels. https://ec.europa.eu/com mission/presscorner/detail/en/ip_20_2484. Accessed 10 May 2022

European Commission (2020c) Proposal for a regulation of the European Parliament and of the Council on on contestable and fair markets in the digital sector (Digital Markets Act). European Commission, Brussels. https://eur-lex.europa.eu/legal-content/en/TXT/?uri=COM%3A2020c% 3A842%3AFIN. Accessed 11 May 2022

European Commission (2020d) Proposal for a regulation of the European Parliament and of the Council on European data governance (Data Governance Act). European Commission, Brussels. https://eur-lex.europa.eu/legal-content/EN/TXT/?uri=CELEX%3A5202 0dPC0767. Accessed 11 May 2022

European Commission (2021) Commission welcomes political agreement to boost data sharing and support European data spaces. Press release, 30 Nov. https://ec.europa.eu/commission/pre sscorner/detail/en/IP_21_6428. Accessed 11 May 2022

European Parliament, Council of the EU (2016) Regulation (EU) 2016/679 of the European Parliament and of the Council of 27 April 2016 on the protection of natural persons with regard to the processing of personal data and on the free movement of such data, and repealing Directive 95/46/EC (General Data Protection Regulation). Official J Eur Union L119(1):1–88. https:// eur-lex.europa.eu/legal-content/EN/TXT/PDF/?uri=CELEX:32016R0679&from=EN. Accessed 1 May 2022

Fadler M, Legner C (2021) Data ownership revisited: clarifying data accountabilities in times of Big Data and analytics. J Bus Anal. https://doi.org/10.1080/2573234X.2021.1945961

Fernandez R (2021) How Big Tech is becoming the government. SOMO, 5 Feb. https://www.somo. nl/how-big-tech-is-becoming-the-government/. 15 May 2022

Fernandez R, Adriaans I, Klinge TJ, Hendrikse R (2020) The financialisation of Big Tech. SOMO (Centre for Research on Multinational Corporations), Amsterdam. https://www.somo.nl/nl/wp-content/uploads/sites/2/2020/12/Engineering_Financial-BigTech.pdf. Accessed 3 May 2022

Fowler J (n.d.) Report: fitness tracker data breach exposed 61 million records and user data online. Website Planet. https://www.websiteplanet.com/blog/gethealth-leak-report/. Accessed 10 May 2022

Garrett PM (2021) "Surveillance capitalism, COVID-19 and social work": a note on uncertain future (s). Br J Soc Work 52(3):1747–1764. https://doi.org/10.1093/bjsw/bcab099

Ghaffary S (2021) Big Tech's employees are one of the biggest checks on its power. Vox-Recode, 29 Dec. https://www.vox.com/recode/22848750/whistleblower-facebook-google-apple-employees. Accessed 3 May

Grunes AP, Stucke ME (2015) No mistake about it: the important role of antitrust in the era of Big Data. Antitrust Source, Online, University of Tennessee Legal Studies Research Paper 269. https://papers.ssrn.com/sol3/Delivery.cfm/SSRN_ID2608540_code869490.pdf?abstractid= 2600051&mirid=1. Accessed 3 May 2022

Harwell D (2022) Facial recognition firm Clearview AI tells investors it's seeking massive expansion beyond law enforcement. The Washington Post, 16 Feb. https://www.washingtonpost.com/techno logy/2022/02/16/clearview-expansion-facial-recognition/. Accessed 10 May 2022

Hummel P, Braun M, Dabrock P (2021) Own data? Ethical reflections on data ownership. Philos Technol 34:545–572. https://doi.org/10.1007/s13347-020-00404-9

Jensen J (2019) Ethical aspects of surveillance capitalism. LinkedIn, 7 Nov. https://www.linkedin. com/pulse/ethical-aspects-surveillance-capitalism-jostein-jensen/. Accessed 15 May 2022

Jolly J (2021) Is Big Tech now just too big to stomach? The Guardian, 6 Feb. https://www.thegua rdian.com/business/2021/feb/06/is-big-tech-now-just-too-big-to-stomach. Accessed 9 May 2022

Jurcys P, Donewald C, Fenwick M et al (2020) Ownership of user-held data: why property law is the right approach. Harv J Law Technol Digest. https://jolt.law.harvard.edu/assets/digestImages/ Paulius-Jurcys-Feb-19-article-PJ.pdf. Accessed 3 May 2022

Kang SS, (2020) Don't blame privacy for Big Tech's monopoly on information. Just Security, 18 Sept. https://www.justsecurity.org/72439/dont-blame-privacy-for-big-techs-monopoly-on-inf ormation/. Accessed 3 May 2022

Kavenna J (2019) Interview. Shoshana Zuboff: 'Surveillance capitalism is an assault on human autonomy'. The Guardian, 4 Oct. https://www.theguardian.com/books/2019/oct/04/shoshana-zuboff-surveillance-capitalism-assault-human-automomy-digital-privacy. Accessed 16 May 2022

Leblond P (2020) How open data could tame Big Tech's power and avoid a breakup. The Conversation, 5 Aug. https://theconversation.com/how-open-data-could-tame-big-techs-power-and-avoid-a-breakup-143962. Accessed 3 May 2022

Lehtiniemi T (2017) Personal data spaces: an intervention in surveillance capitalism? Surveill Soc 15(5):626–639. https://doi.org/10.24908/ss.v15i5.6424

Lin Y (2018) #DeleteFacebook is still feeding the beast—but there are ways to overcome surveillance capitalism. The Conversation, 26 Mar. https://theconversation.com/deletefacebook-is-still-feeding-the-beast-but-there-are-ways-to-overcome-surveillance-capitalism-93874. Accessed 3 May 2022

Lyon D (2019) Surveillance capitalism, surveillance culture and data politics. In: Bigo D, Isin E, Ruppert E (eds) Data politics: worlds, subjects, rights. Routledge, Oxford, pp 64–77

Matsakis L (2019) Break up Big Tech? Some say not so fast. Wired, 7 June. https://www.wired.com/story/break-up-big-tech-antitrust-laws/. Accessed 3 May 2022

Milmo D (2021) Frances Haugen: "I never wanted to be a whistleblower. But lives were in danger". The Observer, 24 Oct. https://www.theguardian.com/technology/2021/oct/24/frances-haugen-i-never-wanted-to-be-a-whistleblower-but-lives-were-in-danger. Accessed 15 May 2022

Moss DL (2019) Breaking up is hard to do: the implications of restructuring and regulating digital technology markets. The Antitrust Source 19(2). https://www.americanbar.org/content/dam/aba/publishing/antitrust-magazine-online/2018-2019/atsource-october2019/oct19_full_source.pdf. Accessed 3 May 2022

Nash, J (2021) Amazon sees profit in your palmprint. Opponents see harmful surveillance capitalism. Biometric Update, 3 Aug. https://www.biometricupdate.com/202108/amazon-sees-profit-in-your-palmprint-opponents-see-harmful-surveillance-capitalism. Accessed 10 May 2022

Nass SJ, Levit LA, Gostin LO (eds) (2009) Beyond the HIPAA privacy rule: enhancing privacy, improving health through research. National Academies Press, Washington DC. https://www.ncbi.nlm.nih.gov/books/NBK9579/. Accessed 10 May 2022

Oremus W (2018) Arc you really the product? The history of a dangerous idea. Slate, 27 Apr. https://slate.com/technology/2018/04/are-you-really-facebooks-product-the-history-of-a-dangerous-idea.html. Accessed 10 May 2022

Perrigo B (2019) "The capabilities are still there." Why Cambridge Analytica whistleblower Christopher Wylie is still worried. Time, 8 Oct. https://time.com/5695252/christopher-wylie-cambridge-analytica-book/. Accessed 15 May 2022

Pistor K (2020) Rule by data: the end of markets? Law Contemp Prob 83:101–124. https://schola rship.law.duke.edu/lcp/vol83/iss2/6. Accessed 4 May 2022

Reuters (2021) Factbox: how Big Tech is faring against U.S. lawsuits and probes. Reuters, 7 Dec. https://www.reuters.com/technology/big-tech-wins-two-battles-fight-with-us-antitrust-enf orcers-2021-06-29/. Accessed 4 May 2022

Rodrigues R, Panagiotopoulos A, Lundgren B et al (2020) SHERPA deliverable 3.3 Report on regulatory options. https://doi.org/10.21253/DMU.11618211.v7

Scroxton A (2021) Mass health tracker data breach has UK impact. Computer Weekly, 14 Sept. https://www.computerweekly.com/news/252506664/Mass-health-tracker-data-breach-has-UK-impact. Accessed 10 May 2022

Simonite T (2021) What really happened when Google ousted Timnit Gebru. Wired, 8 June. https://www.wired.com/story/google-timnit-gebru-ai-what-really-happened/. Accessed 15 May 2022

Sokol DD, Comerford R (2016) Antitrust and regulating Big Data. Geo Mason Law Rev 23:1129–1161. https://papers.ssrn.com/sol3/papers.cfm?abstract_id=2834611. Accessed 4 May 2022

Swartz J (2021) Big Tech heads for "a year of thousands of tiny tech papercuts," but what antitrust efforts could make them bleed? MarketWatch, 27 Dec (updated 1 Jan 2022). https://www.marketwatch.com/story/big-tech-heads-for-a-year-of-thousands-of-tiny-tech-papercuts-but-what-antitrust-efforts-could-make-them-bleed-11640640776. Accessed 4 May 2022

The Economist (2019) Breaking up is hard to do: dismembering Big Tech. The Economist, 24 Oct. https://www.economist.com/business/2019/10/24/dismembering-big-tech. Accessed 4 May 2022

UNESCO (2021) Recommendation on the ethics of artificial intelligence. SHS/BIO/PI/2021/. https://unesdoc.unesco.org/ark:/48223/pf0000381137. Accessed 18 Oct 2022

Van Asbroeck B, Debussche J, César J (2019) Big Data & issues & opportunities: data ownership. Bird & Bird, 25 Mar. https://www.twobirds.com/en/news/articles/2019/global/big-data-and-issues-and-opportunities-data-ownership. Accessed 3 May 2022

Warren E (2019) Here's how we can break up Big Tech. Medium, 8 Mar. https://medium.com/@teamwarren/heres-how-we-can-break-up-big-tech-9ad9e0da324c. Accessed 4 May 2022

Waters R (2019) Three ways that Big Tech could be broken up. Financial Times, 7 June. https://www.ft.com/content/cb8b707c-88ca-11e9-a028-86cea8523dc2. Accessed 4 May 2022

Wolff J, Atallah N (2021) Early GDPR penalties: analysis of implementation and fines through May 2020. J Inf Policy 11(1):63–103. https://doi.org/10.5325/jinfopoli.11.2021.0063

Zuboff S (2015) Big other: surveillance capitalism and the prospects of an information civilization. J Inf Technol 30(1):75–89. https://doi.org/10.1057/jit.2015.5

Zuboff S (2019a) Facebook, Google and a dark age of surveillance capitalism. Financial Times, 25 Jan. https://www.ft.com/content/7fafec06-1ea2-11e9-b126-46fc3ad87c65. Accessed 10 May 2022

Zuboff S (2019b) The age of surveillance capitalism: the fight for a human future at the new frontier of power. Public Affairs, New York

Zuboff S (2020) Surveillance capitalism. Project Syndicate, 3 Jan. https://www.project-syndicate.org/onpoint/surveillance-capitalism-exploiting-behavioral-data-by-shoshana-zuboff-2020-01. Accessed 10 May 2022

Zuboff S (2021) The coup we are not talking about. The New York Times, 29 Jan. https://www.nytimes.com/2021/01/29/opinion/sunday/facebook-surveillance-society-technology.html. Accessed 4 May 2022

Chapter 5
Manipulation

Abstract The concern that artificial intelligence (AI) can be used to manipulate individuals, with undesirable consequences for the manipulated individual as well as society as a whole, plays a key role in the debate on the ethics of AI. This chapter uses the case of the political manipulation of voters and that of the manipulation of vulnerable consumers as studies to explore how AI can contribute to and facilitate manipulation and how such manipulation can be evaluated from an ethical perspective. The chapter presents some proposed ways of dealing with the ethics of manipulation with reference to data protection, privacy and transparency in the of use of data. Manipulation is thus an ethical issue of AI that is closely related to other issues discussed in this book.

Keywords Right to life · Safety · Security · Self-driving cars · Smart homes · Adversarial attacks · Responsibility · Liability · Quality management · Adversarial robustness

5.1 Introduction

In the wake of the 2016 US presidential election and the 2016 Brexit referendum it became clear that AI had been used to target undecided voters and persuade them to vote in a particular direction. Both polls were close, and a change of mind by a single-digit percentage of the voter population would have been enough to change the outcome. It is therefore reasonable to state that these interventions led by artificial intelligence (AI) played a causal role in the ascent of Donald Trump to the American presidency and the success of the Brexit campaign.

These examples of the potential manipulation of elections are probably the most high-profile cases of human action being influenced using AI. They are not the only ones, however, and they point to the possibility of much further-reaching manipulation activities that may be happening already, but are currently undetected.

© The Author(s) 2023
B. C. Stahl et al., *Ethics of Artificial Intelligence*,
SpringerBriefs in Research and Innovation Governance,
https://doi.org/10.1007/978-3-031-17040-9_5

5.2 Cases of AI-Enabled Manipulation

5.2.1 Case 1: Election Manipulation

The 2008 US presidential election has been described as the first that "relied on large-scale analysis of social media data, which was used to improve fundraising efforts and to coordinate volunteers" (Polonski 2017). The increasing availability of large data sets and AI-enabled algorithms led to the recognition of new possibilities of technology use in elections. In the early 2010s, Cambridge Analytica, a voter-profiling company, wanted to become active in the 2014 US midterm election (Rosenberg et al. 2018). The company attracted a $15 million investment from Robert Mercer, a Republican donor, and engaged Stephen Bannon, who later played a key role in President Trump's 2016 campaign and was an important early member of the Trump cabinet. Cambridge Analytica lacked the data required for voter profiling, so it solved this problem with Facebook data (Cadwalladr and Graham-Harrison 2018). Using a permission to harvest data for academic research purposes that Facebook had granted to Aleksandr Kogan, a researcher with links to Cambridge University, the company harvested not just the data of people who had been paid to take a personality quiz, but also that of their friends. This allowed Cambridge Analytica to harvest in total 50 million Facebook profiles, which allowed the delivery of personalised messages to the profile holders and also – importantly – a wider analysis of voter behaviour.

The Cambridge Analytica case led to a broader discussion of the permissible and appropriate uses of technology in Western democracies. Analysing large datasets with a view to classifying demographics into small subsets and tailoring individual messages designed to curry favour with the individuals requires data analytics techniques that are part of the family of technologies typically called AI.

We will return to the question of the ethical evaluation of manipulation below. The questions that are raised by manipulation will become clearer when we look at a second example, this one in the commercial sphere.

5.2.2 Case 2: Pushing Sales During "Prime Vulnerability Moments"

Human beings do not feel and behave the same way all of the time; they have ups and downs, times when they feel more resilient and times when they feel less so. A 2013 marketing study suggests that one can identify typical times when people feel more vulnerable than usual. US women across different demographic categories, for example, have been found to feel least attractive on Mondays, and therefore possibly more open to buying beauty products (PHD Media 2013). This study goes on to suggest that such insights can be used to develop bespoke marketing strategies. While the original study couches this approach in positive terms such as "encourage"

> and "empower", independent observers have suggested that it may be the "grossest
> advertising strategy of all time" (Rosen 2013).

Large internet companies such as Google and Amazon use data they collect about potential customers to promote goods and services that their algorithms suggest searchers are in need of or looking for. This approach could easily be combined with the concept of "prime vulnerability moments", where real-time data analysis is used to identify such moments in much more detail than the initial study.

The potential manipulation described in this second case study is already so widespread that it may not be noticeable any more. Most internet users are used to being targeted in advertising.

The angle of the case that is interesting here is the use of the "prime vulnerability moment", which is not yet a concept widely referred to in AI-driven personal marketing. The absence of a word for this concept does not mean, however, that the underlying approach is not used. As indicated, the company undertaking the original study couched the approach in positive and supportive terms. The outcome of such a marketing strategy may in fact be positive for the target audience. If a person has a vulnerable moment due to fatigue, suggestions of relevant health and wellbeing products might help combat that state. This leads us to the question we will now discuss: whether and in what circumstances manipulation arises, and how it can be evaluated from an ethical position.

5.3 The Ethics of Manipulation

An ethical analysis of the concept of manipulation should start with an acknowledgement that the term carries moral connotations. The Cambridge online dictionary offers the following definition: "controlling someone or something to your own advantage, often unfairly or dishonestly" (Cambridge Dictionary n.d.) and adds that it is used mainly in a disapproving way. The definition thus offers several pointers to why manipulation is seen as ethically problematic. The act of controlling others may be regarded as concerning, especially the fact that it is done for someone's advantage, which is exacerbated if it is done unfairly or dishonestly. In traditional philosophical terms, it is Kant's prominent categorical imperative that prohibits such manipulation on ethical grounds, because one person is being used solely as a means to another person's ends (Kant 1998: 37 [4:428]).

One aspect of the discussion that is pertinent to the first case study is that the manipulation of the electorate through AI can damage democracy.

> AI can have (and likely already has) an adverse impact on democracy, in particular where
> it comes to: (i) social and political discourse, access to information and voter influence, (ii)
> inequality and segregation and (iii) systemic failure or disruption. (Muller 2020: 12)

Manipulation of voters using AI techniques can fall under heading (i) as voter influence. However, it is not clear under which circumstances such influence on voters would be illegitimate. After all, election campaigns explicitly aim to influence voters and doing so is the daily work of politicians. The issue seems to be not so much the fact that voters are influenced, but that this happens without their knowledge and maybe in ways that sidestep their ability to critically reflect on election messages. An added concern is the fact that AI is mostly held and made use of by large companies, and that these are already perceived to have an outsized influence on policy decisions, which can be further extended through their ability to influence voters. This contributes to the "concentration of technological, economic and political power among a few mega corporations [that] could allow them undue influence over governments" (European Parliament 2020: 16).

Another answer to the question why AI-enabled manipulation is ethically problematic is that it is based on privacy infringements and constitutes surveillance. This is certainly a key aspect of the Cambridge Analytica case, where the data of Facebook users was harvested in many cases without their consent or awareness. This interpretation would render the manipulation problem a subproblem of the broader discussion of privacy, data protection and surveillance as discussed in Chap. 3.

However, the issue of manipulation, while potentially linked with privacy and other concerns, seems to point to a different fundamental ethical concern. In being manipulated, the objects of manipulation, whether citizens and voters or consumers, seem to be deprived of their basic freedom to make informed decisions.

Freedom is a well-established ethical value that finds its expressions in many aspects of liberal democracy and forms a basis of human rights. It is also a very complex concept that has been discussed intensively by moral philosophers and others over millennia (Mill 1859; Berlin 2002). While it may sound intuitively plausible to say that manipulating individuals using AI-based tools reduces their freedom to act as they normally would, it is more difficult to determine whether or how this is the case. There are numerous interventions which claim that AI can influence human behaviour (Whittle 2021), for example by understanding cognitive biases and using them to further one's own ends (Maynard 2019). In particular the collecting of data from social media seems to provide a plausible basis for this claim, where manipulation (Mind Matters 2018) is used to increase corporate profits (Yearsley 2017). However, any such interventions look different from other threats to our freedom to act or to decide, such as incarceration and brainwashing.

Facebook users in the Cambridge Analytica case were not forced to vote in a particular way but received input that influenced their voting behaviour. Of course, this is the intended outcome of election campaigns. Clearly the argument cannot be that one should never attempt to influence other people's behaviour. This is what the law and, to some extent, ethics do as a matter of course. Governments, companies and also special interest groups all try to influence, often for good moral reasons. If a government institutes a campaign to limit smoking by displaying gruesome pictures of cancer patients on cigarette packets, then this has the explicit intention of dissuading people from smoking without ostensibly interfering with their basic right to freedom. We mentioned the idea of nudging in Chap. 3, in the context of privacy (Benartzi et al. 2017), which constitutes a similar type of intervention. While nudging is contentious, certainly when done by governments, it is not always and fundamentally unethical.

So perhaps the reference to freedom or liberty as the cause of ethical concerns in the case of manipulation is not fruitful in the discussion of the Cambridge Analytica case. A related alternative that is well established as a mid-level principle from biomedical ethics (Childress and Beauchamp 1979) is that of autonomy. Given that biomedical principles including autonomy have been widely adopted in the AI ethics debate, this may be a more promising starting point. Respect for autonomy is, for example, one of the four ethical principles that the EU's High-Level Expert Group bases its ethics guidelines for trustworthy AI on (AI HLEG 2019). The definition of this principle makes explicit reference to the ability to partake in the democratic process and states that "AI systems should not unjustifiably subordinate, coerce, deceive, manipulate, condition or herd humans" This suggests that manipulation is detrimental to autonomy as it reduces "meaningful opportunity for human choice" (ibid: 12).

This position supports the contention that the problem with manipulation is its detrimental influence on autonomy. A list of requirements for trustworthy AI starts with "human agency and oversight" (ibid: 15). This requirement includes the statement that human autonomy may be threatened when AI systems are "deployed to shape and influence human behaviour through mechanisms that may be difficult to detect, since they may harness sub-conscious processes" (ibid: 17). The core of the problem, then, is that people are not aware of the influence that they are subjected to, rather than the fact that their decisions or actions are influenced in a particular way.

This allows an interesting question to be raised about the first case study (Facebook and Cambridge Analytica). Those targeted were not aware that their data had been harvested from Facebook, but they may have been aware that they were being subjected to attempts to sway their political opinion – or conceivably might have been, if they had read the terms and conditions of Facebook and third-party apps they were using. In this interpretation the problem of manipulation has a close connection to the question of informed consent, a problem that has been highlighted with regard to possible manipulation of Facebook users prior to the Cambridge Analytica event (Flick 2016).

The second case (pushing sales during "prime vulnerability moments") therefore presents an even stronger example of manipulation, because the individuals subjected to AI-enabled interventions may not have been aware of this at all. A key challenge, then, is that technology may be used to fundamentally alter the space of perceived available options, thereby clearly violating autonomy.

Coeckelbergh (2019) uses the metaphor of the theatre, with a director who sets the stage and thereby determines what options are possible in a play. AI can similarly be used to reveal or hide possible options for people in the real world. In this case manipulation would be undetectable by the people who are manipulated, precisely because they do not know that they have further options. It is not always possible to fully answer the question: when does an acceptable attempt to influence someone turn into an unacceptable case of manipulation? But it does point to possible ways of addressing the problem.

5.4 Responses to Manipulation

An ethical evaluation of manipulation is of crucial importance in determining which interventions may be suitable to ensure that AI use is acceptable. If the core of the problem is that political processes are disrupted and power dynamics are affected in an unacceptable manner, then the response could be sought at the political level. This may call for changes to electoral systems or maybe the breaking up of inappropriately powerful large tech companies that threaten existing power balances, as proposed by the US senator and former presidential candidate Warren (2019) and others (Yglesias 2019). Similarly, if the core of the ethical concern is the breach of data protection and privacy, then strengthening or enforcing data protection rules is likely to be the way forward.

While such interventions may be called for, the uniqueness of the ethical issue of manipulation seems to reside in the hidden way in which people are influenced. There are various ways in which this could be addressed. On one hand, one could outlaw certain uses of personal data, for example its use for political persuasion. As political persuasion is neither immoral in principle nor illegal, such an attempt to regulate the use of personal data would likely meet justified resistance and be difficult to define and enforce legally.

A more promising approach would be to increase the transparency of data use. If citizens and consumers understood better how AI technologies are used to shape their views, decisions and actions, they would be in a better position to consciously agree or disagree with these interventions, thereby removing the ethical challenge of manipulation.

Creating such transparency would require work at several levels. At all of these levels, there is the need to understand and explain how AI systems work. Machine learning is currently the most prominent AI application that has given rise to much of the ethical discussion of AI. One of the characteristics of machine learning approaches using neural networks and deep learning (Bengio et al. 2021) is the opacity of the resulting model. A research stream on explainable AI has developed in response to this problem of technical opacity. While it remains a matter of debate whether explainability will benefit AI, or to what degree the internal states of an AI system can be subject to explanation (Gunning et al. 2019), much technical work has been undertaken to provide ways in which humans can make sense of AI and AI outputs. For instance, there have been contributions to the debate highlighting the need for humans to be able to relate to it (Miller 2019; Mittelstadt et al. 2019). Such work could, for example, make it clear to individual voters why they have been selected as targets for a specific political message, or to consumers why they are deemed to be suitable potential customers for a particular product or service.

Technical explainability will not suffice to address the problem. The ubiquity of AI applications means that individuals, even if highly technology-savvy, will not have the time and resources to follow up all AI decisions that affect them and even less to intervene, should these be wrong or inappropriate. There will thus need to be a social and political side to transparency and explainability. This can include the inclusion of stakeholders in the design, development and implementation of AI, which is an intention that one can see in various political AI strategies (Presidency of the Council of the EU 2020; HM Government 2021).

Stakeholder involvement is likely to address some of the problems of opacity, but it is not without problems, as it poses the perennial question: who should have a seat at the table (Borenstein et al. 2021)? It will therefore need to be supplemented with processes that allow for the promotion of meaningful transparency. This requires the creation of conditions where adversarial transparency is possible, for instance where critical civil society groups such as Privacy International[1] are given access to AI systems in order to scrutinise those systems as well as their uses and social consequences. To be successful, this type of social transparency will need a suitable regulatory environment. This may include direct legislation that would force organisations to share data about their systems; a specific regulator with the power to grant access to systems or undertake independent scrutiny; and/or novel standards or processes, such as AI impact assessments, whose findings are required to be published (see Sect. 2.4.1).

5.5 Key Insights

This chapter has shown that concerns about manipulation as an ethical problem arising from AI are closely related to other ethical concerns. Manipulation is directly connected to data protection and privacy. It has links to broader societal structures and the justice of our socio-economic systems and thus relates to the problem of surveillance capitalism. By manipulating humans, AI can reduce their autonomy.

The ethical issue of manipulation can therefore best be seen using the systems-theoretical lens proposed by Stahl (2021, 2022). Manipulation is not a unique feature that arises from particular uses of a specific AI technology; it is a pervasive capability of the AI ecosystem(s). Consequently what is called for is not one particular solution, but rather the array of approaches discussed in this book. In the present chapter we have focused on transparency and explainable AI as key aspects of a successful mitigation strategy. However, these need to be embedded in a larger regulatory framework and are likely to draw on other mitigation proposals ranging from standardisation to ethics-by-design methodologies.

---

[1] https://privacyinternational.org/.

References

AI HLEG (2019) Ethics guidelines for trustworthy AI. High-level expert group on artificial intelligence. European Commission, Brussels. https://ec.europa.eu/newsroom/dae/document.cfm?doc_id=60419. Accessed 25 Sept 2020

Benartzi S, Besears J, Mlikman K et al (2017) Governments are trying to nudge us into better behavior. Is it working? The Washington Post, 11 Aug. https://www.washingtonpost.com/news/wonk/wp/2017/08/11/governments-are-trying-to-nudge-us-into-better-behavior-is-it-working/. Accessed 1 May 2022

Bengio Y, Lecun Y, Hinton G (2021) Deep learning for AI. Commun ACM 64:58–65. https://doi.org/10.1145/3448250

Berlin I (2002) Liberty. Oxford University Press, Oxford

Borenstein J, Grodzinsky FS, Howard A et al (2021) AI ethics: a long history and a recent burst of attention. Computer 54:96–102. https://doi.org/10.1109/MC.2020.3034950

Cadwalladr C, Graham-Harrison E (2018) How Cambridge analytica turned Facebook 'likes' into a lucrative political tool. The Guardian, 17 Mar. https://www.theguardian.com/technology/2018/mar/17/facebook-cambridge-analytica-kogan-data-algorithm. Accessed 1 May 2022

Cambridge Dictionary (n.d.) Manipulation. https://dictionary.cambridge.org/dictionary/english/manipulation. Accessed 11 May 2022

Childress JF, Beauchamp TL (1979) Principles of biomedical ethics. Oxford University Press, New York

Coeckelbergh M (2019) Technology, narrative and performance in the social theatre. In: Kreps D (ed) Understanding digital events: Bergson, Whitehead, and the experience of the digital, 1st edn. Routledge, New York, pp 13–27

European Parliament (2020) The ethics of artificial intelligence: issues and initiatives. European Parliamentary Research Service, Brussels. https://www.europarl.europa.eu/RegData/etudes/STUD/2020/634452/EPRS_STU(2020)634452_EN.pdf. Accessed 1 May 2022

Flick C (2016) Informed consent and the Facebook emotional manipulation study. Res Ethics 12. https://doi.org/10.1177/1747016115599568

Gunning D, Stefik M, Choi J et al (2019) XAI: explainable artificial intelligence. Sci Robot 4(37). https://doi.org/10.1126/scirobotics.aay7120

HM Government (2021) National AI strategy. Office for Artificial Intelligence, London. https://assets.publishing.service.gov.uk/government/uploads/system/uploads/attachment_data/file/1020402/National_AI_Strategy_-_PDF_version.pdf

Kant I (1998) Groundwork of the metaphysics of morals. Cambridge University Press, Cambridge

Maynard A (2019) AI and the art of manipulation. Medium, 18 Nov. https://medium.com/edge-of-innovation/ai-and-the-art-of-manipulation-3834026017d5. Accessed 15 May 2022

Mill JS (1859) On liberty and other essays. Kindle edition, 2010. Digireads.com

Miller T (2019) Explanation in artificial intelligence: insights from the social sciences. Artif Intell 267:1–38. https://doi.org/10.1016/j.artint.2018.07.007

Mind Matters (2018) AI social media could totally manipulate you, 26 Nov. https://mindmatters.ai/2018/11/ai-social-media-could-totally-manipulate-you/. Accessed 15 May 2022

Mittelstadt B, Russell C, Wachter S (2019) Explaining explanations in AI. In: Proceedings of the conference on fairness, accountability, and transparency (FAT*'19). Association for Computing Machinery, New York, pp 279–288. https://doi.org/10.1145/3287560.3287574

Muller C (2020) The impact of artificial intelligence on human rights, democracy and the rule of law. Ad Hoc Committee on Artificial Intelligence (CAHAI), Council of Europe, Strasbourg. https://rm.coe.int/cahai-2020-06-fin-c-muller-the-impact-of-ai-on-human-rights-democracy-/16809ed6da. Accessed 2 May 2022

PHD Media (2013) New beauty study reveals days, times and occasions when U.S. women feel least attractive, 2 Oct. https://www.prnewswire.com/news-releases/new-beauty-study-reveals-days-times-and-occasions-when-us-women-feel-least-attractive-226131921.html. Accessed 11 May 2022

Polonski V (2017) The good, the bad and the ugly uses of machine learning in election campaigns, 30 Aug. Centre for Public Impact, London. https://www.centreforpublicimpact.org/insights/good-bad-ugly-uses-machine-learning-election-campaigns. Accessed 11 May 2022

Presidency of the Council of the EU (2020) Presidency conclusions: the Charter of Fundamental Rights in the context of artificial intelligence and digital change. Council of the European Union, Brussels. https://www.consilium.europa.eu/media/46496/st11481-en20.pdf. Accessed 1 May 2022

Rosen RJ (2013) Is this the grossest advertising strategy of all time? The Atlantic, 3 Oct. https://www.theatlantic.com/technology/archive/2013/10/is-this-the-grossest-advertising-strategy-of-all-time/280242/. Accessed 11 May 2022

Rosenberg M, Confessore N, Cadwalladr C (2018) How Trump consultants exploited the Facebook data of millions. The New York Times, 17 Mar. https://www.nytimes.com/2018/03/17/us/politics/cambridge-analytica-trump-campaign.html. Accessed 11 May 2022

Stahl BC (2021) From computer ethics and the ethics of AI towards an ethics of digital ecosystems. AI Ethics. https://doi.org/10.1007/s43681-021-00080-1

Stahl BC (2022) Responsible innovation ecosystems: ethical implications of the application of the ecosystem concept to artificial intelligence. Int J Inf Manage 62:102441. https://doi.org/10.1016/j.ijinfomgt.2021.102441

Warren E (2019) Here's how we can break up Big Tech. Medium, 8 Mar. https://medium.com/@teamwarren/heres-how-we-can-break-up-big-tech-9ad9e0da324c. Accessed 15 May 2022

Whittle J (2021) AI can now learn to manipulate human behaviour. The Conversation, 11 Feb. https://theconversation.com/ai-can-now-learn-to-manipulate-human-behaviour-155031. Accessed 15 May 2022

Yearsley Y (2017) We need to talk about the power of AI to manipulate humans. MIT Technology Review, 5 June. https://www.technologyreview.com/2017/06/05/105817/we-need-to-talk-about-the-power-of-ai-to-manipulate-humans/. Accessed 15 May 2022

Yglesias M (2019) The push to break up Big Tech, explained. Vox-Recode, 3 May. https://www.vox.com/recode/2019/5/3/18520703/big-tech-break-up-explained. Accessed 15 May 2022

Chapter 6
Right to Life, Liberty and Security of Persons

Abstract Artificial intelligence (AI) can support individuals' enjoyment of life, liberty and security, but it can also have adverse effects on them in a variety of ways. This chapter covers three cases affecting human life, liberty and security: one in transportation (self-driving cars), one in the home (smart security systems) and one in healthcare services (adversarial attacks). The chapter discusses ethical questions and three potential solutions to address AI human rights issues related to life, liberty and security of persons: defining and strengthening liability regimes, implementing quality management systems and adversarial robustness. AI developers, deployers and users must respect the sanctity of human life and embed, value and respect this principle in the design, development and use of their products and/or services. Critically, AI systems should *not* be programmed to kill or injure humans.

Keywords Right to life · Safety · Security · Self-driving cars · Smart homes · Adversarial attacks

6.1 Introduction

All humans enjoy the right to life, liberty and security of the person. The right to life is also included as a core right in 77% of the world's constitutions (UN 2018), is the cornerstone of other rights and is enshrined in international human rights instruments (Table 6.1).

State parties who are signatories to the human rights instruments enshrining the right to life have a duty to take necessary measures to ensure individuals are protected from its violation: i.e. its loss, deprivation or removal.

Artificial intelligence (AI) can support an individual's enjoyment of life, liberty and security by, for example, supporting the diagnosis and treatment of medical conditions. Raso et al. (2018) outline how criminal justice risk assessment tools could benefit low-risk individuals through increased pre-trial releases and shorter sentences. Reports suggest that AI tools could help identify and mitigate human security risks and lower crime rates (Deloitte n.d., Muggah 2017).

© The Author(s) 2023
B. C. Stahl et al., *Ethics of Artificial Intelligence*,
SpringerBriefs in Research and Innovation Governance,
https://doi.org/10.1007/978-3-031-17040-9_6

Table 6.1 Right to life in international human rights instruments

Provision	Human Rights Instrument
Right to life, liberty and security of person	Universal Declaration of Human Rights (UN 1948: art. 3)
Right to life	International Covenant on Civil and Political Rights (UN 1966: art. 6)
Right to life, survival and development	Convention on the Rights of the Child (UN 1989: art. 6)
Right to life	International Convention on the Protection of the Rights of All Migrant Workers and Members of Their Families (UN 1990: art. 9)
Right to life	Convention on the Rights of Persons with Disabilities (UN 2006: art. 10)
Right to life	American Convention on Human Rights (Pact of San José) (UN 1969: art. 4)
Right to life	African Charter on Human and Peoples' Rights (Banjul Charter) (ACHPR 1981: art. 4)
Right to life and dignity in old age	Inter-American Convention on Protecting the Human Rights of Older Persons (OAS 2015: art. 6)
Right to life	European Convention on Human Rights (ECHR 1950: art. 2)

AI can have adverse effects on human life, liberty and security in a variety of ways (Vasic and Billard 2013; Leslie 2019), as elaborated in this chapter. Human rights issues around life, liberty and security of persons are particularly serious, and risks from the use of AI need to be weighed up against the risks incurred when not using AI, in comparison with other innovations. AI systems identified as high-risk (European Commission 2021) include those used in critical infrastructure (e.g. transportation) that could put the life and health of people at risk; in educational or vocational training that determine access to education and the professional course of someone's life (e.g. the scoring of exams); in the safety components of products (e.g. AI applications in robot-assisted surgery); in employment, the management of workers and access to self-employment (e.g. CV-sorting software for recruitment procedures); in essential private and public services (e.g. when credit scoring denies citizens the opportunity to obtain a loan); in law enforcement that may interfere with people's fundamental rights (e.g. evaluation of the reliability of evidence); in migration, asylum and border control management (e.g. verification of the authenticity of travel documents); and in the administration of justice and democratic processes (e.g. applying the law to a concrete set of facts). These categories of high-risk AI systems have the potential to impact the right to life, liberty and security (some in more direct ways than others, but nonetheless relevant).

Life-threatening issues have been raised regarding the use of robot-assisted medical procedures and robotics systems in surgery (Alemzadeh et al. 2016), robot accidents and malfunctions in manufacturing, law enforcement (Boyd 2016), retail and entertainment settings (Jiang and Gainer 1987), security vulnerabilities in smart home hubs (Fránik and Čermák 2020), self-driving and autonomous vehicles (AP and Reuters 2021), and lethal attacks by AI-armed drone swarms and autonomous weapons (Safi 2019). We look at three different cases affecting human life, liberty and security, one in the transportation context (self-driving cars), one related to the home (smart home security), and one in the healthcare service setting (adversarial attacks).

6.2 Cases of AI Adversely Affecting the Right to Life, Liberty and Security of Persons

6.2.1 Case 1: Fatal Crash Involving a Self-driving Car

In May 2016, a Tesla car was the first known self-driving car to be involved in a fatal crash. The 42-year-old passenger/driver died instantly after colliding with a tractor-trailer. The tractor driver was not injured. "According to Tesla's account of the crash, the car's sensor system, against a bright spring sky, failed to distinguish a large white 18-wheel truck and trailer crossing the highway" (Levin and Woolf 2016). An examination by the Florida Highway Patrol concluded that the Tesla driver had not been attentive and had failed to take evasive action. At the same time, the tractor driver had failed, during a left turn, to give right of way, according to the report. (Golson 2017)

In this case, the driver had put his car into Tesla's autopilot mode, which was able to control the car. According to Tesla, its autopilot is "an advanced driver assistance system that enhances safety and convenience behind the wheel" and, "[w]hen used properly" is meant to reduce a driver's "overall workload" (Tesla n.d.). While Tesla clarified that the underlying autonomous software was designed to nudge consumers to keep their hands on the wheels to make sure they were paying attention, that does not seem to have happened in this case and resulted in a fatality. According to Tesla, "the currently enabled Autopilot and Full Self-Driving features require active driver supervision and do not make the vehicle autonomous" (ibid).

In 2018, an Uber test driver in charge of monitoring one of the company's self-driving cars was charged with negligent homicide when it hit and killed a pedestrian. An investigation by the National Transportation Safety Board (NSTB) concluded that the crash had been caused by the Uber test driver being distracted by her phone and implicated Uber's inadequate safety culture (McFarland 2019). The NSTB also found that Uber's system could not correctly classify and predict the path of a pedestrian crossing midblock.

In 2021, two men were killed in Texas after the Tesla vehicle they were in, which was going at a high speed, went off the road and hit a tree. The news report also mentioned that the men been discussing the autopilot feature before they drove off (Pietsch 2021). Evidence is believed to show that no one was driving the vehicle when it crashed.

While drivers seem to expect self-driving cars, as marketed to them, to give them more independence and freedom, self-driving cars are not yet, as stated by Tesla, for example, "autonomous". The autopilot function and the "Full Self-Driving" capability are intended for use with a fully attentive driver with hands on the wheel and ready to take over at any moment.

While some research (Kalra and Groves 2017; Teoh and Kidd 2017) seems to suggest that self-driving cars may be safer than those driven by the average human driver, the main case and the further examples cited here point to human safety challenges from different angles: the safety of the drivers, passengers and other road users (e.g. cyclists, pedestrians and animals) and objects that encounter self-driving cars.

Other standard issues raised about self-driving cars, as outlined by Jansen et al. (2020), relate to *security* (the potential for their hacking leading to the compromising of personal and sensitive data) and *responsibility*, that is, where does responsibility for harms caused lie: with the manufacturer, the system programmer or software engineer, the driver/passenger, or the insurers? A responsibility gap could also occur, as pointed out by the Council of Europe's Committee on Legal Affairs and Human Rights, "where the human in the vehicle—the 'user-in-charge', even if not actually engaged in driving—cannot be held liable for criminal acts and the vehicle itself was operating according to the manufacturer's design and applicable regulations." (Council of Europe 2020). There is also the challenge of shared driving responsibilities between the human driver and the system (BBC News 2020).

The underlying causes that require addressing in these cases include software/system vulnerabilities, inadequate safety risk assessment procedures and oversight of vehicle operators, as well as human error and driver distractions (including a false sense of security) (Clifford Law 2021).

6.2.2 Case 2: Smart Home Hubs Security Vulnerabilities

A smart home hub is a control centre for home automation systems, such as those operating the heating, blinds, lights and internet-enabled electronic appliances. Such systems allow the user to interact remotely with the hub using, for instance, a smartphone. A user who is equipped to activate appliances remotely can arrive at home with the networked gas fire burning and supper ready in the networked oven. However, it is not only the users themselves who can access their smart home hubs, but also external entities, if there are security vulnerabilities, as was the case for three companies operating across Europe. (Fránik and Čermák 2020)

Smart home security vulnerabilities directly affect all aspects of the right to life, liberty and security of the person. E.g., Man-in-the-middle attacks that interrupt or spoof communication between smart home devices and denial-of-service attacks could disrupt or shut devices down and compromise user well-being, safety and security.

Such vulnerabilities and attacks exploiting them can threaten a home, together with the peaceful enjoyment of life and human health within it. Unauthorised access could also result in threats to human life and health. For example, as outlined in a report from the European Union Agency for Cybersecurity (ENISA), safety might be compromised and human life thus endangered by the breach, or loss of control, of a thermostat, a smoke detector, a CO_2 detector or smart locks (Lévy-Bencheton et al. 2015).

When smart home security is exposed to vulnerabilities and threats, these can facilitate criminal actions and intrusions, or could themselves be a form of crime (e.g. physical damage, theft or unauthorised access to smart home assets) (Barnard-Wills et al. 2014).

While there are many other ethical issues that concern smart homes (e.g. access, autonomy, freedom of association, freedom of movement, human touch, informed consent, usability), this case study also further underlines two critical issues connected to the right to life: *security* and *privacy* (Marikyanet al. 2019; Chang et al. 2021). Hackers could spy on people, get access to very personal information and misuse smart-home-connected devices in a harmful manner (Laughlin 2021). Nefarious uses could include the perpetration of identity theft, location tracking, home intrusions and access lock-outs.

The responsibilities for ensuring that smart home devices and services do not suffer from vulnerabilities or attacks are manifold, and lie largely with the manufacturers and service providers, and with users. Users of smart-home-connected devices must carry out their due diligence when purchasing smart devices (by buying from reputable companies with good security track records and ensuring that security is up to the task).

6.2.3 Case 3: Adversarial Attacks in Medical Diagnosis

Medical diagnosis, particularly in radiology, often relies on images. Adversarial attacks on medical image analysis systems are a problem (Bortsova et al. 2021) that can put lives at risk. This applies whether the AI system is tasked with the medical diagnosis or whether the task falls to radiologists, as an experiment with mammogram images has shown. Zhou et al. used a generative adversarial network (GAN) model to make intentional modifications to radiology images taken to detect breast cancer (Zhou et al. 2021). The resulting fake images were then analysed by an AI model and by radiologists. The adversarial samples "fool the AI-CAD model to output a wrong diagnosis on 69.1% of the cases that are initially correctly classified by the AI-CAD model. Five breast imaging radiologists visually identify 29–71% of the adversarial samples" (ibid). In both cases, a wrong cancer diagnosis could lead to risks to health and life.

Adversarial attacks are "advanced techniques to subvert otherwise-reliable machine-learning systems" (Finlayson et al. 2019). These techniques, for example by making tiny image manipulations (adversarial noise) to images that might help confirm a diagnosis, guarantee positive trial results or control the rates of medical interventions to the advantage of those carrying out such attacks (Finlayson et al. 2018).

To raise awareness of adversarial attacks, Rahman et al. (2021) tested COVID-19 deep learning applications and found that they were vulnerable to adversarial example attacks. They report that due to the wide availability of COVID-19 data sets, and because some data sets included both COVID-19 patients' public data and their attributes, they could poison data and launch classified inference attacks. They were able to inject fake audio, images and other types of media into the training data set. Based on this, Rahman et al. (2021) call for further research and the use of appropriate defence mechanisms and safeguards.

The case study and examples mentioned in this section expose the problem of machine and deep learning application vulnerabilities in the healthcare setting. They show that a lack of appropriate defence mechanisms, safeguards and controls would cause serious harm by changing results to detrimental effect.

6.3   Ethical Questions

All the case studies raise several ethical issues. Here we discuss some of the core ones.

6.3.1   Human Safety

To safeguard human safety, which has come to the fore in all three case studies, unwanted harms, risks and vulnerabilities to attack need to be addressed, prevented and eliminated throughout the life cycle of the AI product or service (UNESCO 2021). Human safety is rooted in the value of human life and wellbeing. Safety requires that AI systems and applications should not cause harm through misuse, questionable or defective design and unintended negative consequences. Safety, in the context of AI systems, is connected to ensuring their accuracy, reliability, security and robustness (Leslie 2019). *Accuracy* refers to the ability of an AI system to make correct judgements, predictions, recommendations or decisions based on data or models (AI HLEG 2019). Inaccurate AI predictions may result in serious and adverse effects on human life. *Reliability* refers to the ability of a system to work properly using different inputs and in a range of situations, a feature that is deemed critical for both scrutiny and harms prevention (ibid). *Security* calls for protective measures against vulnerabilities, exploitation and attacks at all levels: data, models, hardware and software (ibid). *Robustness* requires that AI systems use a preventative approach to risk. The systems should behave reliably while minimising unintentional and unexpected harm and preventing unacceptable harm, and at the same time ensuring the physical and mental integrity of humans (ibid).

6.3.2 Privacy

As another responsible AI principle, privacy (see Chap. 3) is also particularly implicated in the first and second case studies. Privacy, while an ethical principle and human right in itself, intersects with the right to life, liberty and security, and supports it with protective mechanisms in the technological context. This principle, in the AI context, includes respect for the privacy, quality and integrity of data, and access to data (AI HLEG 2019). Privacy vulnerabilities manifest themselves in data leakages which are often used in attacks (Denko 2017). Encryption by itself is not seen to provide "adequate privacy protection" (Apthorpe et al. 2017). AI systems must have appropriate levels of security to prevent unauthorised or unlawful processing, accidental loss, destruction or damage (ICO 2020). They must also ensure that privacy and data protection are safeguarded throughout the system's lifecycle, and data access protocols must be in place (AI HLEG 2019). Furthermore, the quality and integrity of data are critical, and processes and data sets used require testing at all stages.

6.3.3 Responsibility and Accountability

When anything goes wrong, we look for *who* is responsible for making decisions about liability and accountability. Responsibility is seen in terms of ownership and/or answerability. In the cases examined here, responsibility might lie with different entities, depending on their role and/or culpability in the harms caused. The cases furthermore suggest that the allocation of responsibility may not be simple or straightforward. In the case of an intentional attack on an AI system, it may be possible to identify the individual orchestrating it. However, in the case of the autonomous vehicle or that of the smart home, the combination of many contributions and the dynamic nature of the system may render the attempt to attribute the actions of the system difficult, if not impossible.

Responsibility lies not only at the point of harm but goes to the point of inception of an AI system. As the ethics guidelines of the European Commission's High-Level Expert Group on Artificial Intelligence outline (AI HLEG 2019), companies must identify the impacts of their AI systems and take steps to mitigate adverse impacts. They must also comply with technical requirements and legal obligations. Where a provider (a natural or legal person) puts a high-risk AI system on the market or into service, they bear the responsibility for it, whether or not they designed or developed it (European Commission 2021).

Responsibility faces many challenges in the socio-technical and AI context (Council of Europe 2019). The first, the challenge of "many hands" (Van de Poel et al. 2012) results as the "development and operation of AI systems typically entails contributions from multiple individuals, organisations, machine components, software algorithms and human users, often in complex and dynamic environments". (Council of Europe 2019). A second challenge relates to how humans placed in the loop are made responsible for harms, despite having only partial control of an AI system, in an attempt by other connected entities to shirk responsibility and liability. A third challenge highlighted is the unpredictable nature of interactions between multiple algorithmic systems that generate novel and potentially catastrophic risks which are difficult to understand (Council of Europe 2019).

For now, responsibility for acts and omissions in relation to an AI product or service and system-related harms lies with humans. The Montreal Declaration for a Responsible Development of AI (2018) states that the development and use of AI "must not contribute to lessening the responsibility of human beings when decisions must be made". However, it also provides that "when damage or harm has been inflicted by an AIS [AI system], and the AIS is proven to be reliable and to have been used as intended, it is not reasonable to place blame on the people involved in its development or use".

Accountability, as outlined by the OECD, refers to.

> the expectation that organisations or individuals will ensure the proper functioning, throughout their lifecycle, of the AI systems that they design, develop, operate or deploy, in accordance with their roles and applicable regulatory frameworks, and for demonstrating this through their actions and decision-making process (for example, by providing documentation on key decisions throughout the AI system lifecycle or conducting or allowing auditing where justified). (OECD n.d.)

Accountability, in the AI context, is linked to auditability (assessment of algorithms, data and design processes), minimisation and reporting of negative impacts, addressing trade-offs and conflicts in a rational and methodological manner within the state of the art, and having accessible redress mechanisms (AI HLEG 2019).

But accountability in the AI context is also not without its challenges, as Busuioc (2021) explains. Algorithm use creates deficits that affect accountability: the compounding of informational problems, the absence of adequate explanation or justification of algorithm functioning (limits on questioning this), and ensuing difficulties with diagnosing failure and securing redress. Various regulatory tools have thus become important to boost AI accountability.

6.4 Responses

Given the above issues and concerns, it is important to put considerable effort into preventing AI human rights issues arising around life, liberty and security of persons, for which the following tools will be particularly helpful.

6.4.1 Defining and Strengthening Liability Regimes

An effective liability regime offers incentives that help reduce risks of harm and provide means to compensate the victims of such harms. "Liability" may be defined by contractual requirements, fault or negligence-based liability, or no-fault or strict liability. With regard to self-driving cars, liability might arise from tort for drivers and insurers and from product liability for manufacturers. Different approaches are adopted to reduce risks depending on the type of product or service.

Are current liability regimes adequate for AI? As of 1 April 2022, there were no AI-specific legal liability regimes in the European Union or United States, though there have been some attempts to define and strengthen existing liability regimes to take into account harms from AI (Karner et al. 2021).

The European Parliament's resolution of 20 October 2020 with recommendations to the European Commission on a civil liability regime for AI (European Parliament 2020) outlined that there was no need for a complete revision of the well-functioning liability regimes in the European Union. However, the capacity for self-learning, the potential autonomy of AI systems and the multitude of actors involved presented a significant challenge to the effectiveness of European Union and national liability framework provisions. The European Parliament recognised that specific and coordinated adjustments to the liability regimes were necessary to compensate persons who suffered harm or property damage, but did not favour giving legal personality to AI systems. It stated that while physical or virtual activities, devices or processes that were driven by AI systems might technically be the direct or indirect cause of harm or damage, this was nearly always the result of someone building, deploying or interfering with the systems (European Parliament 2020). Parliament recognised, though, that the Product Liability Directive (PLD), while applicable to civil liability claims relating to defective AI systems, should be revised (along with an update of the Product Safety Directive) to adapt it to the digital world and address the challenges posed by emerging digital technologies. This would ensure a high level of effective consumer protection and legal certainty for consumers and businesses and minimise high costs and risks for small and medium-sized enterprises and start-ups. The European Commission is taking steps to revise sectoral product legislation (Ragonnaud 2022; Šajn 2022) and undertake initiatives that address liability issues related to new technologies, including AI systems.

A comparative law study on civil liability for artificial intelligence (Karner et al. 2021) questioned whether the liability regimes in European Union Member States provide for an adequate distribution of all risks, and whether victims will be indemnified or remain undercompensated if harmed by the operation of AI technology, even though tort law principles would favour remedying the harm. The study also highlights that there are some strict liabilities in place in all European jurisdictions, but that many AI systems would not fall under such risk-based regimes, leaving victims to pursue compensation via fault liability.

With particular respect to self-driving vehicles, existing legal liability frameworks are being reviewed and new measures have been or are being proposed (e.g. Automated and Electric Vehicles Act 2018; Dentons 2021). These will need to deal with issues that arise from the shifts of control from humans to automated driver assistance systems, and to address conflicts of interest, responsibility gaps (who is responsible and in what conditions, i.e. the human driver/passengers, system operator, insurer or manufacturer) and the remedies applicable.

A mixture of approaches is required to address harms by AI, as different liability approaches serve different purposes: these could include fault- or negligence-based liability, strict liability and contractual liability. The strengthening of provisions for strict liability (liability that arises irrespective of fault or of a defect, malperformance or non-compliance with the law) is highly recommended for high-risk AI products and services (New Technologies Formation 2019), especially where such products and services may cause serious and/or significant and frequent harms, e.g. death, personal injury, financial loss or social unrest (Wendehorst 2020).

6.4.2   Quality Management for AI Systems

Given the risks shown in the case studies presented, it is critical that AI system providers have a good quality management system in place. As outlined in detail in the proposal for the Artificial Intelligence Act (European Commission 2021), this should cover the following aspects:

1. a strategy for regulatory compliance …
2. techniques, procedures and systematic actions to be used for the design, design control and design verification of the high-risk AI system;
3. techniques, procedures and systematic actions to be used for the development, quality control and quality assurance of the high-risk AI system;
4. examination, test and validation procedures to be carried out before, during and after the development of the high-risk AI system, and the frequency with which they have to be carried out,
5. technical specifications, including standards, to be applied and, where the relevant harmonised standards are not applied in full, the means to be used to ensure that the high-risk AI system complies with the requirements set out [in this law];
6. systems and procedures for data management, including data collection, data analysis, data labelling, data storage, data filtration, data mining, data aggregation, data retention and any other operation regarding the data that is performed

before and for the purposes of the placing on the market or putting into service of high-risk AI systems;

7. the risk management system …
8. the setting-up, implementation and maintenance of a post-market monitoring system …
9. procedures related to the reporting of serious incidents and of malfunctioning …
10. the handling of communication with national competent authorities, competent authorities, including sectoral ones, providing or supporting the access to data, notified bodies, other operators, customers or other interested parties;
11. systems and procedures for record keeping of all relevant documentation and information,
12. resource management, including security of supply related measures,
13. an accountability framework setting out the responsibilities of the management and other staff …

6.4.3 Adversarial Robustness

Case 3 demonstrates the need to make AI models more robust to adversarial attacks. As an IBM researcher puts it, "Adversarial robustness refers to a model's ability to resist being fooled" (Chen 2021). This calls for the adoption of various measures, such as the simulation and mitigation of new attacks, via, for example, reverse engineering to recover private data, adversarial training (Tramèr et al. 2018; Bai et al 2021, University of Pittsburgh 2021), using pre-generating adversarial images and teaching the model that these images are manipulated, and designing robust models and algorithms (Dhawale et al. 2022). The onus is clearly on developers to prepare for and anticipate AI model vulnerabilities and threats.

Examples abound of efforts to increase adversarial robustness (Gorsline et al. 2021). Li et al. (2021) have proposed an enhanced defence technique called Attention and Adversarial Logit Pairing (AT + ALP), which, when applied to clean examples and their adversarial counterparts, would help improve accuracy on adversarial examples over adversarial training. Tian et al. (2021) have proposed what they call "detect and suppress the potential outliers" (DSPO), a defence against data poisoning attacks in federated learning scenarios.

6.5 Key Insights

The right to life is the baseline of all rights: the first among other human rights. It is closely related to other human rights, including some that are discussed elsewhere in this book, such as privacy (see Chap. 3) or dignity (see Chap. 7).

In the AI context, this right requires AI developers, deployers and users to respect the sanctity of human life and embed, value and respect this principle in the design, development and use of their products and/or services. Critically, AI systems should *not* be programmed to kill or injure humans.

Where there is a high likelihood of harms being caused, even if accidental, additional precautions must be taken and safeguards set up to avoid them, for example the use of standards, safety-based design, adequate monitoring of the AI system (Anderson 2020), training, and improved accident investigation and reporting (Alemzadeh et al. 2016).

While the technology may have exceeded human expectations, AI must support human life, *not* undermine it. The sanctity of human life must be preserved. What is furthermore required is *sensitivity* to the value of human life, liberty and security. It is insensitivity to harms and impacts that leads to change-resistant problematic actions. Sensitivity requires the ability to understand what is needed and the taking of helpful actions to fulfil that need. It also means remembering that AI can influence, change and damage human life in many ways. This sensitivity is required at all levels: development, deployment and use. It requires continuous learning on the adverse impacts that an AI system may have on human life, liberty and security and avoiding and/or mitigating such impacts to the fullest extent possible.

References

ACHPR (1981) African (Banjul) Charter on Human and Peoples' Rights. Adopted 27 June. African Commission on Human and Peoples' Rights, Banjul. https://www.achpr.org/public/Document/file/English/banjul_charter.pdf. Accessed 24 May 2022

AI HLEG (2019) Ethics guidelines for trustworthy AI. High-Level Expert Group on Artificial Intelligence, European Commission, Brussels. https://ec.europa.eu/newsroom/dae/document.cfm?doc_id=60419. Accessed 25 Sept 2020

Alemzadeh H, Raman J, Leveson N et al (2016) Adverse events in robotic surgery: a retrospective study of 14 years of FDA data. PLoS ONE 11(4):e0151470. https://doi.org/10.1371/journal.pone.0151470

Anderson B (2020) Tesla autopilot blamed on Fatal Japanese Model X crash. Carscoops, 30 April. https://www.carscoops.com/2020/04/tesla-autopilot-blamed-on-fatal-japanese-model-x-crash/. Accessed 24 May 2022

AP, Reuters (2021) US regulators probe deadly Tesla crash in Texas. DW, 19 April. https://p.dw.com/p/3sFbD. Accessed 22 May 2022

Apthorpe NJ, Reisman D, Feamster N (2017) A smart home is no castle: privacy vulnerabilities of encrypted IoT traffic. ArXiv, abs/1705.06805. https://doi.org/10.48550/arXiv.1705.06805

Automated and Electric Vehicles Act (2018) c18. HMSO, London. https://www.legislation.gov.uk/ukpga/2018/18/contents. Accessed 24 May 2022

Bai T, Luo J, Zhao J et al (2021) Recent advances in adversarial training for adversarial robustness. In: Zhou Z-H (ed) Proceedings of the thirtieth international joint conference on artificial intelligence (IJCAI-21), International Joint Conferences on Artificial Intelligence, pp 4312–4321. https://doi.org/10.24963/ijcai.2021/591

Barnard-Wills D, Marinos L, Portesi S (2014). Threat landscape and good practice guide for smart home and converged media. European Union Agency for Network and Information Security (ENISA). https://www.enisa.europa.eu/publications/threat-landscape-for-smart-home-and-media-convergence. Accessed 25 May 2022

BBC News (2020) Uber's self-driving operator charged over fatal crash. 16 September. https://www.bbc.com/news/technology-54175359. Accessed 23 May 2022

Bortsova G, González-Gonzalo C, Wetstein SC et al (2021) Adversarial attack vulnerability of medical image analysis systems: unexplored factors. Med Image Anal 73:102141. https://doi.org/10.1016/j.media.2021.102141

Boyd EB (2016) Is police use of force about to get worse—with robots? POLITICO Magazine, 22 September. https://www.politico.com/magazine/story/2016/09/police-robots-ethics-debate-214273/. Accessed 22 May 2022

Busuioc M (2021) Accountable artificial intelligence: holding algorithms to account. Public Adm Rev 81(5):825–836. https://doi.org/10.1111/puar.13293

Chang V, Wang Z, Xu QA et al (2021). Smart home based on internet of things and ethical issues. In: Proceedings of the 3rd international conference on finance, economics, management and IT business (FEMIB), pp 57–64. https://doi.org/10.5220/0010178100570064

Chen P-Y (2021) Securing AI systems with adversarial robustness. IBM Research. https://research.ibm.com/blog/securing-ai-workflows-with-adversarial-robustness. Accessed 15 May 2022

Clifford Law (2021) The dangers of driverless cars. The National Law Review, 5 May. https://www.natlawreview.com/article/dangers-driverless-cars. Accessed 23 May 2022

Council of Europe (2019) Responsibility and AI: a study of the implications of advanced digital technologies (including AI systems) for the concept of responsibility within a human rights framework. Prepared by the Expert Committee on human rights dimensions of automated data processing and different forms of artificial intelligence (MSI-AUT). https://rm.coe.int/responsability-and-ai-en/168097d9c5. Accessed 25 May 2022

Council of Europe (2020) Legal aspects of "autonomous" vehicles. Report Committee on Legal Affairs and Human Rights, Parliamentary Assembly, Council of Europe. https://assembly.coe.int/LifeRay/JUR/Pdf/DocsAndDecs/2020/AS-JUR-2020-20-EN.pdf. Accessed 25 May 2022

Deloitte (n.d.) Urban future with a purpose: 12 trends shaping the future of cities by 2030. https://www2.deloitte.com/global/en/pages/public-sector/articles/urban-future-with-a-purpose.html.

Denko MW (2017) A privacy vulnerability in smart home IoT devices. Dissertation, University of Michigan-Dearborn. https://deepblue.lib.umich.edu/bitstream/handle/2027.42/139706/49698122_ECE_699_Masters_Thesis_Denko_Michael.pdf. Accessed 25 May 2022

Dentons (2021) Global guide to autonomous vehicles 2021. http://www.thedriverlesscommute.com/wp-content/uploads/2021/02/Global-Guide-to-Autonomous-Vehicles-2021.pdf. Accessed 24 May 2022

Dhawale K, Gupta P, Kumar Jain T (2022) AI approach for autonomous vehicles to defend from adversarial attacks. In: Agarwal B, Rahman A, Patnaik S et al (eds) Proceedings of international conference on intelligent cyber-physical systems. Springer Nature, Singapore, pp 207–221. https://doi.org/10.1007/978-981-16-7136-4_17

ECHR (1950) European Convention on Human Rights. 5 November. European Court of Human Rights, Strasbourg. https://www.echr.coe.int/documents/convention_eng.pdf. Accessed 25 May 2022

European Commission (2021) Proposal for a regulation of the European Parliament and of the Council laying down harmonised rules on artificial intelligence (Artificial Intelligence Act) and amending certain union legislative acts. European Commission, Brussels. https://eur-lex.europa.eu/legal-content/EN/TXT/?uri=CELEX%3A52021PC0206. Accessed 1 May 2022

European Parliament (2020) Resolution of 20 October 2020 with recommendations to the commission on a civil liability regime for artificial intelligence (2020/2014(INL)). https://www.europarl.europa.eu/doceo/document/TA-9-2020-0276_EN.pdf. Accessed 24 May 2022

Finlayson SG, Bowers JD, Ito J et al (2019) Adversarial attacks on medical machine learning. Science 363(6433):1287–1289. https://doi.org/10.1126/science.aaw4399

Finlayson SG, Chung HW, Kohane IS, Beam AL (2018) Adversarial attacks against medical deep learning systems. ArXiv preprint. https://doi.org/10.48550/arXiv.1804.05296

Fránik M, Čermák M (2020) Serious flaws found in multiple smart home hubs: is your device among them? WeLiveSecurity, 22 April. https://www.welivesecurity.com/2020/04/22/serious-flaws-smart-home-hubs-is-your-device-among-them/. Accessed 22 May 2022

Golson J (2017) Read the Florida Highway Patrol's full investigation into the fatal Tesla crash. The Verge, 1 February. https://www.theverge.com/2017/2/1/14458662/tesla-autopilot-crash-acc ident-florida-fatal-highway-patrol-report. Accessed 23 Msay 2022

Gorsline M, Smith J, Merkel C (2021) On the adversarial robustness of quantized neural networks. In: Proceedings of the 2021 Great Lakes symposium on VLSI (GLSVLSI '21), 22–25 June 2021, virtual event. Association for Computing Machinery, New York, pp 189–194. https://doi.org/10. 1145/3453688.3461755

ICO (2020) Guidance on AI and data protection. Information Commissioner's Office, Wilmslow, UK. https://ico.org.uk/for-organisations/guide-to-data-protection/key-dp-themes/gui dance-on-ai-and-data-protection/. Accessed 25 May 2022

Jansen P, Brey P, Fox A et al (2020). SIENNA D4.4: Ethical analysis of AI and robotics technologies V1.https://doi.org/10.5281/zenodo.4068083

Jiang BC, Gainer CA Jr (1987) A cause-and-effect analysis of robot accidents. J Occup Accid 9(1):27–45. https://doi.org/10.1016/0376-6349(87)90023-X

Kalra N, Groves DG (2017) The enemy of good: estimating the cost of waiting for nearly perfect automated vehicles. Rand Corporation, Santa Monica CA

Karner E, Koch BA, Geistfeld MA (2021) Comparative law study on civil liability for artificial intelligence. Directorate-General for Justice and Consumers, European Commission, Brussels. https://data.europa.eu/doi/10.2838/77360. Accessed 24 May 2022

Laughlin A (2021) How a smart home could be at risk from hackers. Which?, 2 July. https://www.which.co.uk/news/article/how-the-smart-home-could-be-at-risk-from-hac kers-akeR18s9eBHU. Accessed 23 May 2022

Leslie D (2019) Understanding artificial intelligence ethics and safety: a guide for the responsible design and implementation of AI systems in the public sector. The Alan Turing Institute. https:// doi.org/10.5281/zenodo.3240529

Levin S, Woolf N (2016) Tesla driver killed while using autopilot was watching Harry Potter, witness says. The Guardian, 1 July. https://www.theguardian.com/technology/2016/jul/01/tesla-driver-killed-autopilot-self-driving-car-harry-potter. Accessed 23 May 2022

Lévy-Bencheton C, Darra E, Tétu G et al (2015) Security and resilience of smart home environments. good practices and recommendations. European Union Agency for Network and Information Security (ENISA). https://www.enisa.europa.eu/publications/security-resilience-good-practices. Accessed 25 May 2022

Li X Goodman D Liu J et al (2021) Improving adversarial robustness via attention and adversarial logit pairing. Front ArtifIntell 4. https://doi.org/10.3389/frai.2021.752831

Marikyan D, Papagiannidis S, Alamanos E (2019) A systematic review of the smart home literature: a user perspective. Technol Forecast Soc Change 138:139–154. https://doi.org/10.1016/j.techfore. 2018.08.015

McFarland M (2019) Feds blame distracted test driver in Uber self-driving car death. CNN Business, 20 November. https://edition.cnn.com/2019/11/19/tech/uber-crash-ntsb/index.html. Accessed 23 May 2022

Montreal Declaration (2018) Montréal declaration for a responsible development of artificial intel-ligence. Université de Montréal, Montreal. https://www.montrealdeclaration-responsibleai.com/ the-declaration. Accessed 21 Sept 2020

Muggah R (2017) What happens when we can predict crimes before they happen? World Economic Forum, 2 February. https://www.weforum.org/agenda/2017/02/what-happens-when-we-can-pre dict-crimes-before-they-happen/. Accessed 16 May 2022

New Technologies Formation (2019) Liability for artificial intelligence and other emerging digital technologies. Expert Group on Liability and New Technologies, Directorate-General for Justice and Consumers, European Commission, Brussels. https://data.europa.eu/doi/10.2838/573689. Accessed 24 May 2022

OAS (2015) Inter-American Convention on Protecting the Human Rights of Older Persons. Forty-fifth regular session of the OAS General Assembly, 15 June. http://www.oas.org/en/sla/dil/docs/inter_american_treaties_A-70_human_rights_older_persons.pdf. Accessed 25 May 2022

OECD (n.d.) Accountability (Principle 1.5). OECD AI Policy Observatory. https://oecd.ai/en/dashboards/ai-principles/P9. Accessed 23 May 2022

Pietsch B (2021) 2 killed in driverless Tesla car crash, officials say. The New York Times, 18 April. https://www.nytimes.com/2021/04/18/business/tesla-fatal-crash-texas.html. Accessed 23 May 2022

Ragonnaud G (2022) Legislative train schedule: revision of the machinery directive (REFIT). European Parliament. https://www.europarl.europa.eu/legislative-train/theme-a-europe-fit-for-the-digital-age/file-revision-of-the-machinery-directive. Accessed 24 May 2022

Rahman A, Hossain MS, Alrajeh NA, Alsolami F (2021) Adversarial examples: security threats to COVID-19 deep learning systems in medical IoT devices. IEEE Internet Things J 8(12):9603–9610. https://doi.org/10.1109/JIOT.2020.3013710

Raso F, Hilligoss H, Krishnamurthy V et al (2018). Artificial intelligence and human rights: opportunities and risks. Berkman Klein Center for Internet and Society Research, Harvard University, Cambridge MA. http://nrs.harvard.edu/urn-3:HUL.InstRepos:38021439. Accessed 25 May 2022

Safi M (2019) Are drone swarms the future of aerial warfare? The Guardian, 4 December. https://www.theguardian.com/news/2019/dec/04/are-drone-swarms-the-future-of-aerial-warfare. Accessed 22 May 2022

Šajn N (2022) Legislative train schedule: general product safety regulation. European Parliament. https://www.europarl.europa.eu/legislative-train/theme-a-new-push-for-european-democracy/file-revision-of-the-general-product-safety-directive. Accessed 24 May 2022

Teoh ER, Kidd DG (2017) Rage against the machine? Google's self-driving cars versus human drivers. J Safety Rs 63:57–60. https://doi.org/10.1016/j.jsr.2017.08.008

Tesla (n.d.) Support: autopilot and full self-driving capability. https://www.tesla.com/support/autopilot. Accessed 23 May 2022

Tian Y, Zhang W, Simpson A. et al (2021). Defending against data poisoning attacks: from distributed learning to federated learning. Computer J bxab192. https://doi.org/10.1093/comjnl/bxab192

Tramèr F, Kurakin A, Papernot N et al (2018) Ensemble adversarial training: attacks and defenses. Paper presented at 6th international conference on learning representations, Vancouver, 30 April – 3 May. https://doi.org/10.48550/arXiv.1705.07204

UN (1948) Universal Declaration of Human Rights. http://www.un.org/en/universal-declaration-human-rights/. Accessed 4 May 2022

UN (1966) International Covenant on Civil and Political Rights. General Assembly resolution 2200A (XXI), 16 December. https://www.ohchr.org/en/instruments-mechanisms/instruments/international-covenant-civil-and-political-rights. Accessed 24 May 2022

UN (1969) American Convention on Human Rights: "Pact of San José, Costa Rica". Signed at San José, Costa Rica, 22 November. https://treaties.un.org/doc/publication/unts/volume%201144/volume-1144-i-17955-english.pdf. Accessed 24 May 2022

UN (1989) Convention on the Rights of the Child. General Assembly resolution 44/25, 20 November. https://www.ohchr.org/en/instruments-mechanisms/instruments/convention-rights-child. Accessed 24 May 2022

UN (1990) International Convention on the Protection of the Rights of All Migrant Workers and Members of Their Families. General Assembly resolution 45/158, 18 December. https://www.ohchr.org/en/instruments-mechanisms/instruments/international-convention-protection-rights-all-migrant-workers. Accessed 24 May 2022

UN (2006) Convention on the Rights of Persons with Disabilities. General Assembly resolution A/RES/61/106, 13 December. https://www.ohchr.org/en/instruments-mechanisms/instruments/convention-rights-persons-disabilities. Accessed 24 May 2005

UN (2018) Universal Declaration of Human Rights at 70: 30 articles on 30 articles – article 3. Press release, 12 November. Office of the High Commissioner for Human Rights,

United Nations. https://www.ohchr.org/en/press-releases/2018/11/universal-declaration-human-rights-70-30-articles-30-articles-article-3. 24 May 2022

UNESCO (2021) Recommendation on the ethics of artificial intelligence. SHS/BIO/REC-AIETHICS/2021. General Conference, 41st, 23 November. https://unesdoc.unesco.org/ark:/48223/pf0000380455. Accessed 25 May 2022

University of Pittsburgh (2021) Cancer-spotting AI and human experts can be fooled by image-tampering attacks. Science Daily, 14 December. https://www.sciencedaily.com/releases/2021/12/211214084541.htm. Accessed 24 May 2022.

Vasic M, Billard A (2013) Safety issues in human-robot interactions. In: Proceedings of the 2013 IEEE international conference on robotics and automation, Karlsruhe, 6–10 May, pp 197–204. https://doi.org/10.1109/ICRA.2013.6630576

Van de Poel I, Fahlquist JN, Doorn N et al (2012) The problem of many hands: climate change as an example. Sci Eng Ethics 18(1):49–67. https://doi.org/10.1007/s11948-011-9276-0

Wendehorst C (2020) Strict liability for AI and other emerging technologies. JETL 11(2):150–180. https://doi.org/10.1515/jetl-2020-0140

Zhou Q, Zuley M, Guo Y et al (2021) (2021) A machine and human reader study on AI diagnosis model safety under attacks of adversarial images. Nat Commun 12:7281. https://doi.org/10.1038/s41467-021-27577-x

Chapter 7
Dignity

Abstract Dignity is a very prominent concept in human rights instruments, in particular constitutions. It is also a concept that has many critics, including those who argue that it is *useless* in ethical debates. How useful or not dignity can be in artificial intelligence (AI) ethics discussions is the question of this chapter. Is it a conversation stopper, or can it help explain or even resolve some of the ethical dilemmas related to AI? The three cases in this chapter deal with groundless dismissal by an automated system, sex robots and care robots. The conclusion argues that it makes perfect sense for human rights proponents to treat dignity as a prime value, which takes precedence over others in the case of extreme dignity violations such as torture, human trafficking, slavery and reproductive manipulation. However, in AI ethics debates, it is better seen as an equal among equals, so that the full spectrum of potential benefits and harms are considered for AI technologies using all relevant ethical values.

Keywords Dignity · AI ethics · Sex robots · Care robots

7.1 Introduction

Most human rights instruments protect the inherent dignity of human beings. For instance, the opening of the Universal Declaration of Human Rights states that "recognition of the inherent dignity and of the equal and inalienable rights of all members of the human family is the foundation of freedom, justice and peace in the world" (UN 1948). And the first article of the German constitution (*Grundgesetz*) reads: "(1) Human dignity shall be inviolable. To respect and protect it shall be the duty of all state authority" (Germany 1949).

Given the focus of this book on artificial intelligence (AI) ethics cases that relate to potential human rights infringements, one might think that the concept of dignity would be very useful. And indeed, as will be seen below, dignity has made an entrance into AI ethics discussions.

However, it is also important to note that the meaning of the term "dignity" is highly contested (Schroeder and Bani-Sadr 2017). The Canadian Supreme Court

© The Author(s) 2023
B. C. Stahl et al., *Ethics of Artificial Intelligence*,
SpringerBriefs in Research and Innovation Governance,
https://doi.org/10.1007/978-3-031-17040-9_7

Fig. 7.1 Dignity classification

even decided that dignity was no longer to be used in anti-discrimination cases as it
was too confusing and difficult to apply.

> [H]uman dignity is an abstract and subjective notion that … cannot only become confusing
> and difficult to apply; it has also proven to be an additional burden on equality claimants.
> (Kapp 2008: 22)

To give specific meaning to the concept of dignity in the context of three AI
ethics cases, the classification model developed by Schroeder and Bani-Sadr (2018)
is summarised below.

Three broad types of dignity can be distinguished: the dignity associated with
specific conduct or roles, the intrinsic dignity of all human beings and the critical
interpretation, which sees dignity as nothing but a slogan to stop debate (ibid: 53)
(see Fig. 7.1).

Aspirational dignity, associated with specific conduct or roles, is not available to
all human beings, and it is possible to distinguish three main varieties.

Dignity as an expression of virtue According to Beyleveld and Brownsword
(2001: 139), Nelson Mandela exemplifies the personification of dignity as virtue. His
fortitude in the face of adversity—throughout decades of imprisonment—deserves
almost universal admiration.

Dignity through rank and position The original, historical meaning of dignity
is related to rank and position within hierarchies. For instance, Machiavelli (2015)
believed that "dignity [is conferred] by antiquity of blood" rather than through actions
individuals can take, such as being virtuous.

Dignity of comportment In Dostoevsky's *Crime and Punishment* (1917), two
impoverished ladies are described whose gloves "were not merely shabby but had
holes in them, and yet this evident poverty gave the two ladies an air of special dignity,
which is always found in people who know how to wear poor clothes." Independently
of virtue or rank, these two ladies display dignified comportment, which Dostoevsky
singles out for praise in the name of dignity.

In contrast to these varieties of aspirational dignity, *intrinsic* dignity is available
to *all* human beings and is described in two main ways within Western philosophy
and Christian thinking.

Dignity as intrinsic worth The most prominent understanding of dignity in
Western philosophy today is based on Immanuel Kant's interpretation of dignity

as intrinsic worth, which is not selective but belongs to all human beings.[1] It cannot be denied even to a vicious man, according to Immanuel Kant (1990: 110 [463]), our translation). Hence, dignity as intrinsic worth is unrelated to virtue and moral conduct. From Kantian-type dignity stems the prohibition against actions that dehumanise and objectify human beings—enshrined in the Universal Declaration of Human Rights (UN 1948)—in the worst cases through slavery, torture or degrading treatment.

Dignity as being created in the image of God The Catholic Church also promotes the idea that *all* human beings have dignity, because all human beings, they maintain, "are created in the image and likeness of God" (Markwell 2005: 1132).

These two groups of dignity interpretations (aspirational and intrinsic) are joined by a third, highly critical position. Ruth Macklin (2003) famously argued that dignity "is a useless concept … [that] can be eliminated [from ethics debates] without any loss of content". Harvard professor Steven Pinker even claimed that dignity is "a squishy, subjective notion" used mostly "to condemn anything that gives someone the creeps" (Pinker 2008). Thinkers in this group believe that dignity is often used as a "conversation stopper" to avoid having an in-depth dialogue about challenging issues (Birnbacher 1996).

How useful dignity might be in AI ethics discussions is the question for this chapter. Is it a conversation stopper, or is it useful in helping understand or even resolve some of the ethical dilemmas related to AI?

7.2 Cases of AI in Potential Conflict with Human Dignity

7.2.1 Case 1: Unfair Dismissal

"Automation can be an asset to a company, but there needs to be a way for humans to take over if the machine makes a mistake," says Ibrahim Diallo (Wakefield 2018). Diallo was jobless for three weeks in 2017 (Diallo 2018) after being dismissed by an automated system for no reason his line manager could ascertain. It started with an inoperable access card, which no longer worked for his Los Angeles office, and led to him being escorted from the building "like a thief" (Wakefield 2018) by security staff following a barrage of system-generated messages. Diallo said the message that made him jobless was "soulless and written in red as it gave orders that dictated my fate. Disable this, disable that, revoke access here, revoke access there, escort out of premises … The system was out for blood and I was its very first victim" (ibid). After three weeks, his line manager identified the problem (an employee who had left the company had omitted to approve an action) and reinstated Mr Diallo's contractual rights.

---

[1] One of us has dealt elsewhere with the challenge that Immanuel Kant bestows dignity only upon *rational* beings (Schroeder and Bani-Sadr 2017).

Commenting on the case, AI expert Dave Coplin noted: "It's another example of a failure of human thinking where they allow it to be humans versus machines rather than humans plus machines" (ibid). Another commentator used the term "dignity" in connection with this case, noting that "the dignity of human beings and their 'diminishing value' [is at stake] as we approach the confluence of efficiencies gained from the increasing implementation of artificial intelligence and robotics" (Diallo 2018).

Being escorted from the building like a thief—that is, a criminal—as Diallo put it, and without any wrongdoing on his part, can indeed be interpreted in terms of a loss of dignity. Psychological research has shown that being wrongly accused of criminal offences can have severe consequences for the accused, including for their sense of self and their sense of dignity.

> Along with changes in personality, participants also experienced various other losses related to their sense of self, for example loss of *dignity* and credibility … and loss of hope and purpose for the future [italics added]. (Brooks and Greenberg 2020)

Interpreted this way, the dignity lost by Diallo would be aspirational dignity, dignity that is conduct-related. Being publicly suspected of unlawful or immoral behaviour can, then, lead to a sense of having lost dignity. However, the reason why dignity is not helpful in this AI ethics case is that it is unnecessary to the argument. A dignity interpretation does not move the case forward. There is no moral dilemma to be solved. It is obvious that a human being should not be treated like a criminal and made redundant for the sole reason that an opaque system is unresponsive—in this example, to Diallo's line manager. AI designed to assist with human resources decisions should be understandable, and, as Dave Coplin has noted, should operate on the basis of humans *plus* machines, not humans *versus* machines.

This problem is not unique to AI, a fact that can easily be verified with reference to the success rates of unfair dismissal cases brought by employees. Useful figures are available from Australia, for instance, which radically reformed its dismissal regulations in 2006. Freyens and Oslington (2021) put the success rate of employees claiming unfair dismissal at 47–48% for the period from 2001 to 2015 (a time when cases are unlikely to have been influenced significantly by AI decision-making). Hence, almost half of the employees who challenged employers about their dismissal were deemed to have been unfairly dismissed, as Diallo was, yet most likely without the involvement of AI systems.

The challenge is summarised by Goodman and Kilgallan (2021) from the point of view of employers:

> AI will continue to develop and will likely outperform humans in some aspects of working life. However, the technology has wide implications and employers should be cautious. AI has two notable flaws: the human error of thinking AI is infallible and the lack of transparency in its outcomes.

From the point of view of employees like Ibrahim Diallo, it is also vital for the system that made him jobless to significantly increase its transparency.

7.2.2 Case 2: Sex Robots

> One of the first sex robots available to buy is called Harmony (Boran 2018). The female body of this sex doll is combined with a robotic head, which can turn; the mouth can smile and the eyes can blink. The robot's AI element is steered through an app on the owner's phone. While Harmony cannot stand up, conversation is possible as the app stores information about the owner. The machine doll has been described as "a more sophisticated, sexy Alexa" built on "a very reductive stereotype of femininity: narrow waist, big breasts, curvaceous hips, long blonde hair" (Cosmopolitan 2021). Harmony and other sex robots have been at the centre of highly controversial ethics debates.

The concept of dignity plays a big role in ethics debates about sex robots. Sex robots have been promoted as a way of "achieving 'dignity' … [by enabling] physical touch, intimacy, and sexual pleasure … [for] disabled people" and for "men and women rejected sexually by other men, or women" (Zardiashvili and Fosch-Villaronga 2020). However, it is also argued that "the everlasting availability as well as the possibility to perform any sexual activity violates gender dignity and equality, causing harm that is understood as objectification and commodification" (Rigotti 2020).

The type of dignity both of these discourses appeal to is *intrinsic dignity*, which requires all human beings to be treated with respect and not objectified. Or, as Immanuel Kant (1998: 110 [4:428]) put it in his principle of humanity,

the human being … *exists* as an end in itself, *not merely as a means* to be used by this or that will at its discretion; instead [the human being] … must … always be regarded *at the same time as an end.*

Two main routes are available for discussing the potential moral dilemma of sex robots and their impact on the intrinsic dignity of human beings.

The first route links human rights to sexual rights and sexual autonomy. This route can start, for instance, with the World Health Organization's (WHO) strategy to count "sexual … well-being" as "fundamental to the overall health and well-being of individuals" and therefore related to article 25 of the Universal Declaration of Human Rights (WHO n.d.b) (see box).

> **Universal Declaration of Human Rights, Article 25**
>
> "Everyone has the right to a standard of living adequate for the health and well-being of himself and of his family." (UN 1948)

Historically, the sexual rights movement focused on the prevention of harm, in particular emphasising the rights of girls and women to be free from sexual violence.

Later, the same movement focused on the rights to self-expression of sexual inclination without fear of discrimination for lesbian, gay and transgender people (Miller 2009).

In 2002, going beyond the focus on harm and discrimination, a WHO-commissioned report defined sexual rights as "human rights that are already recognized in national laws, international human rights documents and other consensus statements. They include the right of all persons … to … pursue a satisfying, safe and pleasurable sexual life" (WHO n.d.a). This sexual right to a pleasurable sexual life is limited by the injunction not to infringe the rights of others (ibid). The step from this type of sexual autonomy to sex robots is short.

> If sexual freedom is an integral part of personal autonomy and interference is illegitimate whenever consent and private acts are involved, then robotic sexual intercourse will take place at home, coming to no direct harm to others and falling within the buyer's right to privacy. (Rigotti 2019)

Despite some highly optimistic predictions about the benefits of sex robots—e.g. they will allegedly fill the void in the lives of people who have no-one and therefore provide a "terrific service" to humankind (Wiseman 2015)—the linchpin in this discussion will be the consideration of potential harm. And this is the starting point of the second route for discussing the potential moral dilemma of sex robots and their impact on the intrinsic dignity of human beings.

This second route assumes as its *starting* point that the harm from sex robots is inevitable. For instance, Professor Kathleen Richardson's Campaign Against Sex Robots warns against "reinforcing female dehumanisation" and seeks "to defend the dignity and humanity of women and girls" (CASR n.d.). In her view, sex robots will strengthen a relationship model where buyers of sex can turn off empathy towards the sellers of sex and cement relations of power where one party is not recognised as a human being, but simply as a needs fulfiller framed according to male desires (Richardson 2015): a sole means, not simultaneously an end in herself, in Kantian terminology.

Richardson's views are strengthened by the fact that sex robots are almost exclusively female (Rigotti 2019), that one can speak of "a market by men for men" (Cosmopolitan 2021) and that "the tech's development is largely … focused on the fulfillment of straight male desire" (Edwards 2016). Sex robots are thus seen as an extension of sex work (Richardson uses the term "prostitution"), where "the buyer of sex is at liberty to ignore the state of the other person as a human subject who is turned into a thing" (Richardson 2015).

At this point, most commentators will note that sex robots *are* things, they do not have to be objectified *into* things. Somewhat cynically, one could ask: what harm does it do if a fanatical Scarlett Johansson admirer builds a sex robot in her image, which winks and smiles at him (Pascoe 2017)? It's only a thing, and it might even do some good. For instance, it has been suggested that sex robots, used within controlled environments, could help redirect "the sexual behavior of high-risk child molesters … without endangering real children" (Zara et al. 2022).

Fig. 7.2 Golden mean on sex robot positions?

The two routes to discussing the ethical issues of sex robots sketched above could also be illustrated as a continuum with two opposing poles which meet where sexual autonomy that does not create harm might be the Aristotelian golden mean (Aristotle 2000: 102 [1138b]). (Aristotle argued that one should always strive to find the middle between excess and deficiency: courage, for instance, lying between recklessness and cowardice (Aristotle 2000: 49 [1115b]) (see Fig. 7.2).

Ensuring that sex robots are used without creating harm might pose serious challenges, which are discussed further below. First, we look at care robots.

7.2.3   Case 3: Care Robots

> An old lady sits alone in her sheltered accommodation stroking her pet robot seal. She has not had any human visitors for days. A humanoid robot enters the room, delivers a tray of food, and leaves after attempting some conversation about the weather, and encouraging her to eat it all up. The old lady sighs, and reluctantly complies with the robot's suggestions. When she finishes eating, she goes back to stroking the pet robot seal: 'At least you give my life some meaning' she says. (Sharkey 2014)

Amanda Sharkey paints this picture of an old lady and her care robot in her paper "Robots and Human Dignity: A Consideration of the Effects of Robot Care on the Dignity of Older People".

Despite the emphasis on dignity in the title, she then delivers a very even-handed risk and benefit analysis of care robots, with only limited reference to potential dignity violations. In particular, she distinguishes three types of care robots for the elderly: first, assistive robots, which can, for instance, help with feeding and bathing, or moving a person with limited mobility from a bed to a wheelchair; second, monitoring robots, which can, for instance, detect falls, manage diaries or provide reminders about taking medication; and third, companion robots, which often take the form of pets such as the robot seal in the case description.

In her discussion of the ethical challenges of care robots, Sharkey establishes only three connections with dignity. *Assistive* robots can increase human dignity for elderly people, she believes, in particular by increasing mobility and access to social interaction. *Monitoring* robots, likewise, can increase human dignity, she argues, by enabling elderly people to live independently for longer than otherwise possible. The only ethical challenge related to human dignity Sharkey identifies is related to *companion* robots, which could infantilise elderly people in the eyes of carers or undermine their self-respect, if offered as a sole replacement for human interaction.

Arguably, the concept of dignity does no important work in the first two cases. If a technology can increase mobility and access to social interaction, and enable elderly people to live independently for longer than otherwise possible, the positive contributions to wellbeing are easily understood without reference to the contested concept of dignity (Stahl and Coeckelberg 2016). A similar point could be made using research by Robinson et al. (2014):

> [O]lder people indicated that they would like a robot for detecting falls, controlling appli-
> ances, cleaning, medication alerts, making calls and monitoring location. Most of these tasks
> point towards maintaining independence and dignity.

If one removed the term "dignity" in this quote, which is a technique suggested by dignity critics to ascertain whether the concept is useful or not (Macklin 2003), then the potential benefit of robotic technology in elderly care remains, namely maintaining independence.

Perhaps unsurprisingly, then, when Stahl and Coeckelbergh (2016) examine exactly the same problem (the ethical challenges of health and care robots) they make no reference to dignity at all. Where they align with Sharkey, but use different wording, is on the question of "cold and mechanical" machine care, which might be seen as abandoning elderly people and handing them "over to robots devoid of human contact" (ibid). Stahl and Coeckelberg (ibid) ask whether this might be an objectification of care receivers. In other words, they employ the Kantian concept of dignity, also used by Richardson to justify her Campaign Against Sex Robots, to ask whether elderly people who are cared for by robots are objectified—in other words, turned from subjects into things.

7.3   Ethical Questions Concerning AI and Dignity

There is something behind all three AI cases that is not easy to describe and warrants ethical attention. It is linked to how people are seen by others and how this relates to their own self-respect. Jean-Paul Sartre (1958: 222) used the concept of the *gaze* to describe this situation.

> [T]he Other is the indispensable mediator between myself and me. ... By the mere appearance
> of the Other, I am put in the position of passing judgement on myself as an object, for it is
> as an object that I appear to the other.

It was the fact that Diallo was seen by others and possibly judged as a wrongdoer that made his forced removal from his workplace a potential dignity issue. The concern that sex robots may reinforce human relationships that see women and girls as mere needs fulfillers of male sexual desires and not as human subjects in their own right is what seems to drive Richardson's campaign to ban sex robots. And it is the fact that elderly people in care who engage with their robot pets are possibly seen as infantilised in the eyes of carers that is one of Sharkey's main concerns.

The judgement-filled gaze of the other and the link to self-respect as one passes judgement on oneself through the eyes of the other is what makes the three AI cases above relevant to dignity debates. "Dignity" therefore seems a suitable word to describe at least some of the moral dilemmas involved in the cases. One might hence argue that dignity is not a mere conversation stopper, but a helpful concept in the context of AI ethics. Let us examine this further.

It is suggested above that the concept of dignity is not necessary in drawing useful ethical conclusions from the first case, that of unfair dismissal due to an opaque AI system. This position would maintain that there is no moral dilemma as there are no proponents of competing claims who have to find common ground. Unfair dismissal due to opaque AI ought to be avoided.[2] Any technical or organisational measures (e.g. ethics by design, see Sect. 2.4.2) that can reasonably be used, should be used to achieve this goal. At the same time, it is essential that remedies be available to employees and workers who find themselves in a situation similar to Diallo's. Thus, a report presented to the UK Trades Union Congress (Allen and Masters 2021: 77) stressed the following:

> Unfair dismissal legislation should protect employees ... from dismissal decisions that are factually inaccurate or opaque in the usual way. The use of AI-powered tools to support such decisions does not make any difference to this important legal protection.

Cases 2 and 3 are more complex in terms of their dignity angle, especially Case 2, sex robots. We shall first discuss Case 3 to provide additional leads for Case 2.

In "'Oh, Dignity Too?' Said the Robot: Human Dignity as the Basis for the Governance of Robotics", Zardiashvili and Fosch-Villaronga (2020) identify eight major ethical concerns in employing care robots for the elderly: safe human–robot interaction, the allocation of responsibility, privacy and data protection loss, autonomy restriction, deception and infantilisation, objectification and loss of control, human–human interaction decrease, and long-term consequences. While the authors recognise that "[r]obots might be the solution to bridge the loneliness that the elderly often experience; they may help wheelchair users walk again, or may help navigate the blind" (ibid), they also "acknowledge that human contact is an essential aspect of personal care and that ... robots for healthcare applications can challenge the dignity of users" (ibid).

We have themed seven of the above concerns into five headings and removed the eighth as it applies to all emerging technologies ("Technology ... may have long-term consequences that might be difficult to foresee"). For our interpretation of Zardiashvili and Fosch-Villaronga's (2020) ethical concerns in relation to care robots and the elderly see Fig. 7.3.

---

[2] Unfair dismissal due to performance being measured by inaccurate algorithms is not covered here, as it is more relevant to Chap. 2 on discrimination than this one on dignity. For relevant literature, see De Stefano (2018).

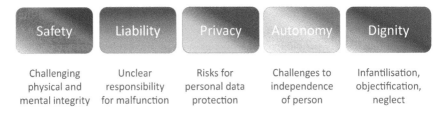

Safety	Liability	Privacy	Autonomy	Dignity
Challenging physical and mental integrity	Unclear responsibility for malfunction	Risks for personal data protection	Challenges to independence of person	Infantilisation, objectification, neglect

Fig. 7.3 Ethical concerns about care robots and the elderly

As Fig. 7.3 shows, dignity is only one of five main ethical concerns identified in relation to care robots and the elderly. This is not unusual. It is often the case that several ethical values or principles need to be protected to achieve an optimal ethical outcome. These values may even conflict. For instance, it may be that the safety of a device could be increased through a more invasive collection of personal data. Stakeholders then have to weigh up the relative importance of safety versus the relative importance of privacy. It is in this weighing-up process that dignity may be an outlier value.

As the founding principle of many constitutions around the world, dignity is often given precedence over all other values. For instance, in the Daschner case (Schroeder 2006), the German constitutional court ruled in 2004 that a threat of duress by police forces to extract information was an unacceptable violation of the dignity of a detainee, even though the police forces were as certain as they could be that the child kidnapper in their custody was the only person who could reveal the whereabouts of an 11-year old who might be starving in an undetected location.[3] The presiding judge noted: "Human dignity is inviolable. Nobody must be made into an object, a bundle of fear" (Rückert 2004, our translation). The inviolability of the value of dignity meant it took precedence over all other values, including what the police forces thought was the right to life of a child.

If the power given to dignity in constitutional courts were to colour other moral debates, such as those on care robots and the elderly, important ethical factors might be ignored. Dignity could then indeed become a conversation stopper, overriding safety, privacy, autonomy and liability issues and moving discussions away from potential benefits. It has already been noted by Sharkey (2014) that (assistive) care robots can increase the wellbeing of elderly people by improving mobility and access to social interaction. Likewise, (monitoring) care robots can increase human well-being by enabling elderly people to live independently for longer than otherwise possible.

In the case of care robots, potential dignity issues therefore have to be treated as one of several ethical challenges, without being given privileged importance. Only then can the ethical risks for the technology be addressed proportionally in the context of serious social care staff shortages. Taking the UK as an example, around ten per cent of social care posts were vacant in 2020 and an additional need for 650,000 to

---

[3] In fact, the child had already been killed by the kidnapper.

950,000 new adult social care jobs is anticipated by 2035 (Macdonald 2020). It is in this context that the development of care robots for the elderly might become an ethical goal in its own right.

As a geriatric nurse commented, "Care robots don't substitute for the human being—they help when no one else is there to help" (Wachsmuth 2018).

Two aspects of the care robot discussion are useful when turning to sex robots. First, unless dealing with extreme dignity violations, such as torture, human trafficking, slavery or reproductive manipulation (Bourcarde 2004), the ethical value of dignity should be treated as an equal among equals and not as an ethical value that automatically overrides all others. For instance, those who demand a ban on sex robots to promote the dignity of women and girls (Richardson 2015) seem to make a categorical claim without considering the ethical value of sexual autonomy. This position has been criticised from a range of angles, e.g. that it relies on unjustified parallels to sex work or that it ignores the demand for male sex dolls and toys (Hancock 2020).

The position we want to take here seeks to achieve the Aristotelian mean between two extremes that would, on the one hand, ban sex robots and, on the other, declare them an important service to humankind (see Fig. 7.2). Instead, the right to "pursue a satisfying, safe and pleasurable sexual life" (WHO n.d.a) may include the use of sex robots to foster sexual autonomy, while researchers and regulators should monitor both potential harms *and* potential benefits.

Benefits could, for instance, involve improving "the satisfaction of the sexual needs of a user" (Fosch-Villaronga and Poulsen 2020) who might have difficulties accessing alternatives, thereby contributing to their health and wellbeing.

Harms could range from very practical considerations, such as sexually transmitted infections from sex robots employed by multiple users in commercial sex work settings (Hancock 2020), to seeking responses to actions that might be criminalised if a human person were involved. One of the most controversially discussed topics in this regard is child sex robots for use by paedophiles.

> In July 2014, the roboticist Ronald Arkin suggested that child sex robots could be used to treat those with paedophilic predilections in the same way that methadone is used to treat heroin addicts. … But most people seem to disagree with this idea, with legal authorities in both the UK and US taking steps to outlaw such devices. (Danaher 2019b)

Using the UK example, since 2017, the Crown Prosecution Service has outlawed the import of child sex dolls, referring to the 1876 Customs Consolidation Act, which forbids the importation of obscene items (Danaher 2019a). This approach is applicable to child sex robots, but other legal avenues have been suggested, in particular using the UK child protection framework or the UK's 2003 Sexual Offences Act to forbid the use of child sex robots (Chatterjee 2020). On the other hand, "proponents of love and sex with robots would argue that a CSB [childlike sexbot] could have a twofold interest: protecting children from sexual predators and by the same token, treating the latter" (Behrendt 2018). One solution between the two extremes is to restrict the use of child sex robots to cases requiring medical authorisation and under strict medical supervision (ibid).

It is noteworthy that the concept of harm rather than that of dignity is usually evoked in the case of child sex robots, which also leads to our conclusion. All ethical aspects that would normally be considered with an emerging technology have to be considered in the process so that dignity is not used as a conversation stopper in assessing this case.

> Robot technology may have moral implications, contribute to the loss of human contact, reinforce existing socio-economic inequalities or fail in delivering good care. (Fosch-Villaronga and Poulsen 2020)

We therefore argue that the use of sex robots should not be ruled out categorically based on dignity claims alone.

7.4 Key Insights

From the early days of drafting human rights instruments, dignity seems to be the concept that has succeeded in achieving consensus between highly diverse negotiators (Schroeder 2012). One negotiator may interpret dignity from a religious perspective, another from a philosophical perspective and yet another from a pragmatic perspective (Tiedemann 2006). This is possible because dignity does not seem to be ideologically fixed in its meaning, and thus allows a basic consensus between different world views.

This advantage could, however, become a problem in AI ethics debates that are concerned with human rights, if dignity considerations are given the power to override all other ethical values. While this would make perfect sense to human rights proponents in the case of extreme dignity violations such as torture, human trafficking, slavery and reproductive manipulation (Bourcarde 2004), dignity is better seen as an equal among equals in AI ethics debates, especially given the risk of losing it altogether, an approach recommended by those who believe dignity is a mere slogan.

The dignity of the elderly is an important consideration in the design and employment of care robots, but it should not be a conversation stopper in the case of sex robots. As with other ethical dilemmas, the full spectrum of potential benefits and harms needs to be considered using all relevant ethical values. In the case of sex robots this can range from the empowerment of "persons with disabilities and older adults to exercise their sexual rights, which are too often disregarded in society" (Fosch-Villaronga and Poulsen 2020) to restrictive regulation for sex robots that enable behaviour that would be criminal in sex work (Danaher 2019b).

References

Allen R, Masters D (2021) Technology managing people: the legal implications. Trades Union Congress, London. https://www.tuc.org.uk/sites/default/files/Technology_Managing_People_2021_Report_AW_0.pdf. Accessed 13 May 2022

Aristotle (2000) Nicomachean Ethics (trans: Crisp R). Cambridge University Press, Cambridge

Behrendt M (2018) Reflections on moral challenges posed by a therapeutic childlike sexbot. In: Cheok AD, Levy D (eds) Love and sex with robots. Springer Nature Switzerland, Cham, pp 96–113. https://doi.org/10.1007/978-3-319-76369-9_8

Beyleveld D, Brownsword R (2001) Human dignity in bioethics and biolaw. Oxford University Press, Oxford

Birnbacher D (1996) Ambiguities in the concept of Menschenwürde. In: Bayertz K (ed) Sanctity of life and human dignity. Kluwer Academic, Dordrecht, pp 107–121. https://doi.org/10.1007/978-94-009-1590-9_7

Boran M (2018) Robot love: the race to create the ultimate AI sex partner. The Irish Times, 1 November. https://www.irishtimes.com/business/technology/robot-love-the-race-to-create-the-ultimate-ai-sex-partner-1.3674387. Accessed 13 May 2022

Bourcarde K (2004) Folter im Rechtsstaat? Die Bundesrepublik nach dem Entführungsfall Jakob von Metzler. Self-published, Gießen, Germany. http://www.bourcarde.eu/texte/folter_im_rechtsstaat.pdf. Accessed 16 June 2021

Brooks SK, Greenberg N (2020) Psychological impact of being wrongfully accused of criminal offences: a systematic literature review. Med Sci Law 61(1): 44–54. https://doi.org/10.1177/0025802420949069

CASR (n.d.) Our story. Campaign Against Sex Robots. https://campaignagainstsexrobots.org/our-story/. Accessed 13 May 2022

Chatterjee BB (2020) Child sex dolls and robots: challenging the boundaries of the child protection framework. Int Rev Law Comput Technol 34(1):22–43. https://doi.org/10.1080/13600869.2019.1600870

Cosmopolitan (2021) Sex robots: how do sex robots work and can you buy a sex robot? Cosmopolitan, 12 July. https://www.cosmopolitan.com/uk/love-sex/sex/a36480612/sex-robots/. Accessed 13 May 2022

Danaher J (2019a) How should we regulate child sex robots: restriction or experimentation? 4 February (blog). BMJ Sex Reprod Health. https://blogs.bmj.com/bmjsrh/2020/02/04/child-sex-robots/. Accessed 13 May 2022

Danaher J (2019b) Regulating child sex robots: restriction or experimentation? Med Law Rev 27(4):553–575. https://doi.org/10.1093/medlaw/fwz002

De Stefano V (2018) "Negotiating the algorithm": automation, artificial intelligence and labour protection. Employment Working Paper No 246. International Labour Office, Geneva. https://www.ilo.org/wcmsp5/groups/public/---ed_emp/---emp_policy/documents/publication/wcms_634157.pdf. Accessed 14 May 2022

Diallo I (2018) The machine fired me: no human could do a thing about it! https://idiallo.com/blog/when-a-machine-fired-me. Accessed 12 May 2022

Dostoevsky F (1917) Crime and punishment (trans: Garnett C). PF Collier & Son, New York. http://www.bartleby.com/318/32.html. Accessed 12 May 2022

Edwards S (2016) Are sex robots unethical or just unimaginative as hell? Jezebel, 7 April. https://jezebel.com/are-sex-robots-unethical-or-just-unimaginative-as-hell-1769358748. 13 May 2022

Fosch-Villaronga E, Poulsen A (2020) Sex care robots: exploring the potential use of sexual robot technologies for disabled and elder care. Paladyn 11:1–18.https://doi.org/10.1515/pjbr-2020-0001

Freyens BP, Oslington P (2021) The impact of unfair dismissal regulation: evidence from an Australian natural experiment. Labour 35(2):264–290. https://doi.org/10.1111/labr.12193

Germany (1949) Basic Law for the Federal Republic of Germany. Federal Ministry of Justice and Federal Office of Justice, Berlin. https://www.gesetze-im-internet.de/englisch_gg/englisch_gg. pdf. Accessed 12 May 2022

Goodman T, Kilgallon P (2021) The risks of using AI in employment processes. People Management, 28 September. https://www.peoplemanagement.co.uk/article/1741566/the-risks-using-ai-employment-processes. Accessed 13 May 2022

Hancock E (2020) Should society accept sex robots? Changing my perspective on sex robots through researching the future of intimacy. Paladyn 11:428–442. https://doi.org/10.1515/pjbr-2020-0025

Kant I (1990) Metaphysische Anfangsgründe der Tugendlehre. Felix Meiner Verlag, Hamburg

Kant I (1998) Groundwork of the metaphysics of morals. Cambridge University Press, Cambridge MA

Kapp RV (2008) 2 SCR 483. https://scc-csc.lexum.com/scc-csc/scc-csc/en/item/5696/index.do. Accessed 12 May 2022

Macdonald M (2020) The health and social care workforce gap. Insight, 10 January. House of Commons Library, London. https://commonslibrary.parliament.uk/the-health-and-social-care-workforce-gap/. Accessed 13 May 2022

Machiavelli N (2015) The prince (trans: Marriott WK). Wisehouse Classics, Sweden

Macklin R (2003) Dignity is a useless concept. BMJ 327:1419–1420. https://doi.org/10.1136/bmj. 327.7429.1419

Markwell H (2005) End-of-life: a Catholic view. Lancet 366:1132–1135. https://doi.org/10.1016/ s0140-6736(05)67425-9

Miller A (2009) Sexuality and human rights: discussion paper. International Council on Human Rights Policy, Versoix. https://biblioteca.cejamericas.org/bitstream/handle/2015/654/Sexuality_ Human_Rights.pdf. Accessed 13 May 2022

Pascoe A (2017) This Scarlett Johansson Robot is uncomfortably realistic. Marie Claire Australia, 10 May. https://www.marieclaire.com.au/scarlett-johansson-robot-sex-doll. Accessed 13 May 2022

Pinker S (2008) The stupidity of dignity. The New Republic, 28 May. https://newrepublic.com/art icle/64674/the-stupidity-dignity. Accessed 12 May 2022

Richardson K (2015) The asymmetrical 'relationship': parallels between prostitution and the development of sex robots. ACM SIGCAS Comput Soc 45(3):290–293. https://doi.org/10.1145/287 4239.2874281

Rigotti C (2020) How to apply Asimov's first law to sex robots. Paladyn 11:161–170. https://doi. org/10.1515/pjbr-2020-0032

Rigotti C (2019) Sex robots: a human rights discourse? OpenGlobalRights, 2 May. https://www. openglobalrights.org/sex-robots-a-human-rights-discourse/. Accessed 13 May 2022

Robinson H, MacDonald B, Broadbent E (2014) The role of healthcare robots for older people at home: a review. Int J of Soc Robot 6:575–591. https://doi.org/10.1007/s12369-014-0242-2

Rückert S (2004) Straflos schuldig. Die Zeit, 22 December. https://www.zeit.de/2004/53/01____ Leiter_2. Accessed 13 May 2022

Sartre J-P (1958) Being and nothingness (trans: Barnes HE). Methuen & Co, London

Schroeder D (2006) A child's life or a "little bit of torture"? State-sanctioned violence and dignity. Camb Q Healthc Ethics 15(2):188–201. https://doi.org/10.1017/S0963180106060233

Schroeder D (2012) Human rights and human dignity. Ethical Theory Moral Pract 15:323–335. https://doi.org/10.1007/s10677-011-9326-3

Schroeder D, Bani-Sadr A-H (2017) Dignity in the 21st century: middle east and west. Springer Int Publishin AG, Cham. https://doi.org/10.1007/978-3-319-58020-3

Sharkey A (2014) Robots and human dignity: a consideration of the effects of robot care on the dignity of older people. Ethics Inf Technol 16:63. https://doi.org/10.1007/s10676-014-9338-5

Stahl BC, Coeckelberg M (2016) Ethics of healthcare robotics: towards responsible research and innovation. Robot Auton Syst 86:152–161. https://doi.org/10.1016/j.robot.2016.08.018

Tiedemann P (2006) Was ist Menschenwürde? Wissenschaftliche Buchgesellschaft, Darmstadt

UN (1948) Universal declaration of human rights. http://www.un.org/en/universal-declaration-human-rights/. Accessed 4 May 2022

Wachsmuth I (2018) Robots like me: challenges and ethical issues in aged care. Front Psychol 9:432. https://doi.org/10.3389/fpsyg.2018.00432

Wakefield J (2018) The man who was fired by a machine. BBC News, 21 June. https://www.bbc.com/news/technology-44561838. Accessed 12 May 2022

WHO (n.d.b) Sexual health: overview. World Health Organization. https://www.who.int/health-topics/sexual-health#tab=tab_1. Accessed 13 May 2022

WHO (n.d.a) Sexual and reproductive health: gender and human rights. World Health Organization. https://www.who.int/reproductivehealth/topics/gender_rights/sexual_health/en/. Accessed 25 November 2021

Wiseman E (2015) Sex, love and robots: is this the end of intimacy? The Observer, 13 December. https://www.theguardian.com/technology/2015/dec/13/sex-love-and-robots-the-end-of-intimacy. Accessed 13 May 2022

Zara G, Veggi S, Farrington DP (2022) Int J Soc Robot 14:479–498. https://doi.org/10.1007/s12369-021-00797-3

Zardiashvili L, Fosch-Villaronga E (2020) "Oh, dignity too?" said the robot: human dignity as the basis for the governance of robotics. Minds Mach 30:121–143. https://doi.org/10.1007/s11023-019-09514-6

Chapter 8
AI for Good and the SDGs

Abstract In 2015, 193 nations came together to agree Agenda 2030: 17 goals ranging from the elimination of poverty to the building of partnerships to achieve those goals. The spirit of the UN Sustainable Development Goals (SDGs) is to leave no one behind. Artificial intelligence (AI) has a great potential to assist in reaching the SDGs. For instance, using algorithms on new and vast agricultural data sets can improve the efficiency of agriculture practices and thereby contribute to SDG 1, "Zero hunger". However, the high energy consumption, computational resources and levels of expertise required for AI can exacerbate existing inequalities. At the same time, potentially useful AI applications such as seasonal climate forecasting have led to the accelerated laying off of workers in Peru and credit denial to poor farmers in Zimbabwe and Brazil. If AI for Good is to be truly realised, AI's potential to worsen inequality, to overexploit resources, to be undertaken through "helicopter research" and to focus on SDG issues relevant mainly to high-income countries must be overcome, ideally in close collaboration and engagement with potential beneficiaries in resource-limited settings.

Keywords AI for good · AI ethics · SDGs · Helicopter research

8.1   Introduction

Artificial intelligence (AI) is one of the few emerging technologies that are very prominently linked to the UN Sustainable Development Goals (SDGs) (UN n.d.a). Through 17 goals and 169 targets, 193 nations resolved "to end poverty and hunger everywhere; to combat inequalities within and among countries; [and] to build peaceful, just and inclusive societies" by 2030 (UN 2015). One could perhaps even argue that AI has been linked directly to international justice and sustainability through the SDGs.

"AI for Good" is a UN-led digital platform[1] that identifies AI solutions to problems relevant to the SDGs. The site offers mostly information about big data sets

---

[1] https://ai4good.org/.

© The Author(s) 2023
B. C. Stahl et al., *Ethics of Artificial Intelligence*,
SpringerBriefs in Research and Innovation Governance,
https://doi.org/10.1007/978-3-031-17040-9_8

provided as links to other sites. For instance, data sets on primary energy production and consumption as well as renewable energy data are provided in relation to SDG 7, "Affordable and clean energy". More unusual links from the AI for Good platform include AI-generated photographs designed to increase empathy with distant strangers. To achieve this increase in empathy, AI calculations transformed pictures of a Boston neighbourhood into images reminiscent of a war-ravaged Syrian city. The results of this DeepEmpathy project were linked to the AI for Good site under SDG 1, "Zero poverty" (Scalable Cooperation n.d.).

A similar initiative, AI for SDGs,[2] led by the Chinese Academy of Sciences, also collates projects from around the world, mapped onto individual SDGs. For instance, an Irish project using remote sensing data, Microsoft Geo AI Data Science Virtual Machines and GIS mapping "to develop machine learning models that can identify agricultural practices" leading to a decline of bees was linked to SDG 2, "Zero hunger", and SDG 15, "Life on land" (AI for SDGs Think Tank 2019).

Many philosophers and ethicists make a distinction between doing no harm and doing good. Prominently, Immanuel Kant distinguished the two by referring to perfect and imperfect duties. According to Kant, certain actions such as lying, stealing and making false promises can never be justified, and it is a perfect duty not to commit those acts (Kant 1965: 14f [397f]). Even without understanding the complicated Kantian justification for perfect duties (categorical imperatives, ibid 42f [421f]), one should find complying with this ethical requirement straightforward, by doing no intentional harm. Imperfect duties, on the other hand, are more difficult to comply with, as they are open-ended. Kant also calls them virtue-based (Kant 1990: 28f [394f]). How much help to offer to the needy is a typical example. Until one has exhausted one's own resources? Or by giving 10% of one's own wealth?

This Kantian distinction is also prominent in the law and everyday morality; as "in both law and ordinary moral reasoning, the avoidance of harm has priority over the provision of benefit" (Keating 2018). AI for Good would then fall into the second category, providing benefits, and by implication become an area of ethics and morality, which is more difficult to assess.

Both case studies are examples of trying to provide benefits. However, as they are drawn from the real world, they blur the lines of the Kantian distinctions. The first case study also illustrates direct harm to vulnerable populations, and the second illustrates a high likelihood of potential harm and a lack of care and equity in international collaboration.

While the intentionality of harm is decisive for Kant when assessing moral actions, lack of due diligence or care has long been identified as a shortcoming in ethical action (Bonnitcha and McCorquodale 2017). (Kant famously said that there is nothing in the world that is ethical or good per se other than a good will—Kant 1965: 10 [393].) Similarly, the 2000-year-old "*Primum non nocere, secundum cavere, tertium sanare*" (Do no harm, then act cautiously, then heal) (Weckbecker 2018) has been employed in the twenty-first century to describe responsible leadership in business and innovation (Leisinger 2018: 120–122). Technologies can often be used for purposes that were not

---

[2] https://ai-for-sdgs.academy/.

originally foreseen or intended, which is why responsiveness and care are required in responsible innovation today (Owen et al. 2013: 35).

8.2 Cases of AI for Good or Not?

"Farming 4.0", "precision agriculture" and "precision farming" (Auernhammer 2001) are all terms used to express, among other things, the employment of big data and AI in agriculture. The International Society of Precision Agriculture has defined precision agriculture as follows:

> Precision agriculture is a management strategy that gathers, processes and analyzes temporal, spatial and individual data and combines it with other information to support management decisions according to estimated variability for improved resource use efficiency, productivity, quality, profitability and sustainability of agricultural production. (ISPA n.d.)

Precision agriculture even has its own academic journal, which covers topics from machine learning methods for crop yield prediction (Burdett and Wellen 2022) to neural networks for irrigation management (Jimenez et al. 2021).

In the service of precision agriculture, AI is useful, for example, in processing vast data amounts for weather forecasting, climate monitoring and decadal predictions (climate predictions of up to a decade) with the ultimate aim of increasing forecast quality (Dewitte et al. 2021). Examples of the benefits of increased forecast quality could be earlier evacuation in the case of severe weather incidents such as tornados and reduced irrigation if future rainfall could be forecast with high precision.

8.2.1 Case 1: Seasonal Climate Forecasting in Resource-Limited Settings

Seasonal climate forecasting (SCF) is used to predict severe weather, such as droughts and floods, in order to provide policymakers and farmers with the means to address problems in an anticipatory rather than a reactive manner (Klemm and McPherson 2017). Lemos and Dilling (2007) have argued that the benefits of SCF mostly reach those "that are already more resilient, or more resource-rich … in terms of … ability to cope with hazards and disasters". By contrast, those who are most at risk of being pushed below the poverty line by severe weather have been harmed in cases in Zimbabwe, Brazil and Peru. In Zimbabwe and Brazil, poor farmers were denied credit after SCF results predicted a drought (ibid). In Zimbabwe, "misinterpretation of the probabilistic nature of the forecast by the banking sector" might have played a role in decision-making about credits (Hammer et al. 2001). SCF forecasting in Peru also led to accelerated layoffs of workers in the fishing industry due to "a forecast of El Niño and the prospect of a weak season." (Lemos and Dilling 2007)

Agenda 2030, the underlying framework for the SDGs, makes the following commitment: "Leave no one behind" (UNSDG n.d.). The case above shows that some of those most in need of not being left behind have suffered as a result of new seasonal climate forecasting techniques.

SDG 9 focuses on fostering innovation and notes in its first target that affordable and equitable access to innovation *for all* should be aimed for (UN n.d.a). While the above cases from Zimbabwe, Brazil and Peru precede the SDGs, the potential for AI to "exacerbate inequality" has since been identified as a major concern for Agenda 2030 (Vinuesa et al. 2020). We will return to this problem after the second case.

8.2.2   Case 2: "Helicopter Research"

> In 2014, a research team from higher-income countries requested access to vast amounts of mobile phone data from users in Sierra Leone, Guinea and Liberia to track population movements during the Ebola crisis. They argued that the value of such data was undeniable in the public health context of the Ebola crisis (Wesolowski 2014). Other researchers disagreed and maintained that quantified population movements would not reveal how the Ebola virus spread (Maxmen 2019). As no ethics guidelines on providing access to mobile phone data existed in Sierra Leone, Guinea and Liberia, government time was spent deliberating whether to provide such access. This time expended on debating access rights, it was argued, "could have been better spent handling the escalating crisis" (ibid). Liberia decided to deny access owing to privacy concerns (ibid) and the research was undertaken on mobile phone data from Sierra Leone. The data showed that fewer people travelled during the Ebola travel ban, but it did not assist in tracking Ebola. (ibid)

One could analyse this case ethically from a harm perspective too, if valuable government time was indeed lost that could have been used to handle the Ebola crisis, as one case commentator argued. One could also analyse it in the context of *potential* harm from privacy breaches when researchers obtain big data sets from countries that have limited means to ensure privacy, especially during a crisis. So-called data-for-good projects have "analysed calls from tens of millions of phone owners in Pakistan, Bangladesh, Kenya and at least two dozen other low- and middle-income nations" (Maxmen 2019) and it has been argued that

> concerns are rising over the lack of consent involved; the potential for breaches of privacy, even from anonymized data sets; and the possibility of misuse by commercial or government entities interested in surveillance. (ibid)

However, we will analyse the case from the perspective of "helicopter research", defined thus:

> The practice of Global North … researchers making roundtrips to the Global South … to collect materials and then process, analyze, and publish results with little to no involvement from local collaborators is referred to as "helicopter research" or "parachute research". (Haelewaters et al. 2021)

Helicopter research thrives in crisis. For instance, during the same 2014 Ebola crisis a social scientist from the North collected social science data without obtaining ethics approval for his research, taking undue advantage of the fragile national regulatory framework for overseeing research (Tegli 2017). Before the publication of his results, the researcher realised that he would need research ethics approval to publish. He had already left the country and asked a research assistant to make the case for retrospective approval. The approval was denied by the relevant research ethics committee (ibid).

One of the main problems of helicopter research is the lack of involvement of local researchers, potentially leading to colonial assumptions about what will help another country best. Often benefits for researchers from the Global North are clear (e.g. access to data, publications, research grants), while benefits might not materialise at all locally, in the Global South (Schroeder et al. 2021). We will return to this in the next section, but here, in the context of obtaining large-scale phone data during a crisis, we can cite a news feature in Nature reporting that

> researchers … say they have witnessed the roll-out of too many technological experiments during crises that don't help the people who most need it. … [A] digital-governance researcher … cautions that crises can be used as an excuse to rapidly analyse call records without frameworks first being used to evaluate their worth or to assess potential harms. (Maxmen 2019)

8.3 Ethical Questions Concerning AI for Good and the SDGs

At first sight, AI for Good seems to deserve celebration, especially when linked to the SDGs. And it is likely that praise is warranted for many efforts, possibly most (Caine 2020). However, the spectre of inequities and unintended harm due to helicopter research or a lack of due diligence looms large. AI for Good may be reminiscent of other efforts where technological solutions have been given precedence over alternatives and where local collaborators have not been consulted, or have even been excluded from contributing.

Another similarly named movement is called GM for Good,[3] and examples of helicopter research on the application of genetically modified (GM) technologies in resource-limited settings are not hard to find.

In 2014, a US university aimed to produce a transgenic banana containing beta-carotene to address vitamin A deficiency in Uganda. Later the research was abandoned for ethical reasons. During the human food trials conducted among US-based students, safety issues and undue inducement concerns materialised. However, the study also raised concerns in Uganda, in particular about the potential release of a transgenic fruit, the risks of undermining local food and cultural systems, and the risks of reducing banana agrobiodiversity. Uganda is home to non-modified banana

---

[3] https://gm4good.org/.

varieties that are already higher in beta-carotene than the proposed transgenic variety. Uninvited intrusions into local food systems that were not matched to local needs were unwelcome and considered inappropriate (Van Niekerk and Wynberg 2018).

Analysing the problems of building GM solutions for populations on the poverty line, Kettenburg et al. (2018) made the following suggestion in the context of Golden Rice, another contentious example (Kettenburg et al. 2018):

> To transcend the reductionism of regarding rice as mere nutrient provider, neglecting its place in the eco- and cultural system … and of describing vitamin A-deficient populations as passive victims … we propose to reframe the question: from "how do we create a rice plant producing beta-carotene?" … to "how do we foster the well-being of people affected by malnutrition, both in short and long terms?"

AI for Good can also be susceptible to the weaknesses of helicopter research and reductionism for the following five reasons.

8.3.1 The Data Desert or the Uneven Distribution of Data Availability

AI relies on data. Machine learning and neural networks are only possible with the input of data. Data is also a highly valuable resource (see Chap. 4 on surveillance capitalism). In this context, a South African report speaks of the "data desert", with worrying figures such as that statistical capacity has *decreased* over the past 15 years in 11 out of 48 African countries (University of Pretoria 2018: 31). This is highly relevant to the use of AI in the context of SDGs. For instance, Case 1 used the records of mobile phone calls during a crisis to track population movements. "However, vulnerable populations are less likely to have access to mobile devices" (Rosman and Carman 2021).

The data desert has at least two implications. First, if local capacity is not available to generate a sufficient amount of data for AI applications in resource-limited settings, it might have to be generated by outsiders, for example researchers from the Global North "helicoptering" into the region. Second, such helicopter research has then the potential to increase the digital divide, as local capacities are left undeveloped. (See below for more on the digital divide.) In this context, Shamika N Sirimanne, Director of Technology and Logistics for the UN Conference on Trade and Development, says, "As the digital economy grows, a data-related divide is compounding the digital divide" (UNCTAD 2021).

8.3.2 The Application of Double Standards

Helicopter research can in effect be research that is *only* carried out in lower-income settings, as it would not be permitted, or would be severely restricted, in higher-income settings, for instance due to the potential for privacy breaches from the large-scale processing of mobile phone records. For example, there is no evidence in the literature of any phone tracking research having been used during the catastrophic 2021 German floods in the Ahr district, even though almost 200 people died and it took weeks to track all the deceased and all the survivors (Fitzgerald et al. 2021). One could speculate that it would have been very hard to obtain consent to gather mobile phone data, even anonymised data, for research from a German population, even in a crisis setting.

8.3.3 Ignoring the Social Determinants of the Problems the SDGs Try to Solve

SDG 2 "Zero hunger" refers to a long-standing problem that Nobel economics laureate Amartya Sen ascribed to entitlement failure rather than a shortage of food availability (Sen 1983). He used the Bengal famine of 1943 to show that the region had more food in 1943 than in 1941, when no famine was experienced. To simplify the argument, the first case study above could be called a study of how the social determinants of hunger were ignored. By trying to improve the forecasting of severe weather in order to give policymakers and farmers options for action in anticipation of failed crops, SCF overlooks the fact that this information, in the hands of banks and employers, could make matters even worse for small-scale farmers and seasonal labourers. That is because the latter have no resilience or resources for addressing food shortages (Lemos and Dilling 2007), the social determinants of hunger.

Another example. An AI application has been developed that identifies potential candidates for pre-exposure prophylaxis in the case of HIV (Marcus et al. 2020). Pre-exposure prophylaxis refers to the intake of medication to prevent infection with HIV. However, those who might need the prophylaxis the most can experience major adherence problems related to SDG 2 "Zero hunger", such as this patient explained.

> When you take these drugs, you feel so hungry. They are so powerful. If you take them on an empty stomach they just burn. I found that sometimes I would just skip the drugs, but not tell anyone. These are some of the things that make it difficult to survive. (Nagata et al. 2012)

An AI solution on its own, without reference to the social determinants of health such as local food security, might therefore not succeed for the most vulnerable segments of populations in resource-limited settings. The type of reductionism attributed to Golden Rice and the Uganda banana scenario described above is likely to occur as well when AI for Good researchers tackle SDGs without local collaborators, which leads to yet another challenge, taking Africa as an example.

8.3.4   The Elephant in the Room: The Digital Divide and the Shortage of AI Talent

> AI depends on high quality broadband. This creates an obvious problem for Africa: given the continent's many connectivity challenges, people must be brought online before they can fully leverage the benefits of AI. (University of Pretoria 2018: 27)

Only an estimated 10% of Africans in rural areas have access to the internet, a figure that goes up to just 22% for urban populations (ibid). These figures are dramatic enough, but the ability to develop AI is another matter altogether. Analysing the potential of AI to contribute to achieving the SDGs, a United Nations Development Programme (UNDP) publication notes that the "chronic shortage of talent able to improve AI capabilities, improve models, and implement solutions" is a critical bottleneck (Chui et al. 2019).

Chronic shortage of AI talent is a worldwide challenge, even for large commercial set-ups. For instance, DeLoitte has commented that "companies [in the US] across all industries have been scrambling to secure top AI talent from a pool that's not growing fast enough." (Jarvis 2020) Potential new staff with AI capabilities are even lured from universities before completing their degrees to fill the shortage (Kamil 2019).

At the same time, a partnership such as 2030Vision, whose focus is the potential for AI to contribute to achieving the SDGs, is clear about what that requires.

> Training significantly more people in the development and use of AI is essential … We need to ensure we are training and supporting people to apply AI to the SDGs, as these fields are less lucrative than more commercially-oriented sectors (e.g. defense and medicine). (2030Vision 2019: 17)

Yet even universities in high-income countries are struggling to educate the next generation of AI specialists. In the context of the shortage of AI talent, one university executive speaks of "a 'missing generation' of academics who would normally teach students and be the creative force behind research projects", but who are now working outside of the university sector (Kamil 2019).

To avoid the potential reductionism of helicopter research in AI for Good, local collaborators are essential, yet these need to be competent local collaborators who are trained in the technology. This is a significant challenge for AI, owing to the serious shortage of workers, never mind trainers.

8.3.5   Wider Unresolved Challenges Where AI and the SDGs Are in Conflict

Taking an even broader perspective, AI and other information and communication technologies (ICTs) might challenge rather than support the achievement of SDG 13, which focuses on climate change. Estimates of electricity needs suggest that "up to

20% of the global electricity demand by 2030" might be taken up by AI and other ICTs, a much higher figure than today's 1% (Vinuesa et al. 2020).

An article in Nature argues that AI-powered technologies have a great potential to create wealth, yet argues that this wealth "may go mainly to those already well-off and educated while … others [are left] worse off" (ibid). The five challenges facing AI for Good that we have enumerated above must be seen in this context.

Preventing helicopter research and unintentional harm to vulnerable populations in resource-limited settings is one of the main aims of the Global Code of Conduct for Research in Resource-Poor Settings (TRUST 2018) (see also Schroeder 2019). Close collaboration with local partners and communities throughout all research phases is its key ingredient. As we went to press, the journal Nature followed major funders (e.g. the European Commission) and adopted the code in an effort "to improve inclusion and ethics in global research collaborations" and "to dismantle systemic legacies of exclusion" (Nature 2022).

8.4 Key Insights

Efforts by AI for Good are contributing to the achievement of Agenda 2030 against the background of a major digital divide and a shortage of AI talent, potentially leading to helicopter research that is not tailored to local needs. This digital divide is just one small phenomenon characteristic of a world that distributes its opportunities extremely unequally. According to Jeffrey Sachs, "there is enough in the world for everyone to live free of poverty and it won't require a big effort on the part of big countries to help poor ones" (Xinhua 2018). But this help cannot be dispensed colonial-style to be effective; it has to be delivered in equitable collaborations with local partners and potential beneficiaries.

What all the challenges facing AI for Good described in this chapter have in common is the lack of equitable partnerships between those who are seeking solutions for the SDGs and those who are meant to benefit from the solutions. The small-scale farmers and seasonal workers whose livelihoods are endangered as a result of the application of seasonal climate forecasting, as well as the populations whose mobile phone data are used without proper privacy governance, are meant to be beneficiaries of AI for Good activities, yet they are not.

A saying usually attributed to Mahatma Gandhi expresses it this way: "Whatever you do for me but without me, you do against me." To make AI for Good truly good for the SDGs, AI's potential to "exacerbate inequality", its potential for the "over-exploitation of resources" and its focus on "SDG issues that are mainly relevant in those nations where most AI researchers live and work" (Vinuesa et al. 2020) must be monitored and counteracted, ideally in close collaboration and engagement with potential beneficiaries in resource-limited settings.

References

AI for SDGsThink Tank (2019) Curbing the decline of wild and managed bees. International Research Center for AI Ethics and Governance, Institute of Automation, Chinese Academy of Sciences. https://ai-for-sdgs.academy/case/151. Accessed 19 May 2022

Auernhammer H (2001) Precision farming: the environmental challenge. Comput Electron Agric 30(1–3):31–43. https://doi.org/10.1016/S0168-1699(00)00153-8

Bonnitcha J, McCorquodale R (2017) The concept of 'due diligence' in the UN guiding principles on business and human rights. Eur J Int Law 28(3):899–919. https://doi.org/10.1093/ejil/chx042

Burdett H, Wellen C (2022) Statistical and machine learning methods for crop yield prediction in the context of precision agriculture. Precision Agric. https://doi.org/10.1007/s11119-022-098 97-0

Caine M (2020) This is how AI could feed the world's hungry while sustaining the planet. World Economic Forum, 24 September. https://www.weforum.org/agenda/2020/09/this-is-how-ai-could-feed-the-world-s-hungry-while-sustaining-the-planet/. Accessed 20 May 2022

Chui M, Chung R, Van Heteren, A (2019) Using AI to help achieve Sustainable Development Goals. United Nations Development Programme, 21 January. https://www.undp.org/blog/using-ai-help-achieve-sustainable-development-goals. Accessed 20 May 2022

Dewitte S, Cornelis JP, Müller R, Munteanu A (2021) Artificial intelligence revolutionises weather forecast, climate monitoring and decadal prediction. Remote Sens 13(16):3209. https://doi.org/10.3390/rs13163209

Fitzgerald M, Angerer C, Smith P (2021) Almost 200 dead, many still missing after floods as Germany counts devastating cost. NBC News, 19 July. https://www.nbcnews.com/news/world/almost-200-dead-many-still-missing-after-floods-germany-counts-n1274330. Accessed 20 May 2022

Haelewaters D, Hofmann TA, Romero-Olivares AL (2021) Ten simple rules for Global North researchers to stop perpetuating helicopter research in the Global South. PLoS Comput Biol 17(8):e1009277. https://doi.org/10.1371/journal.pcbi.1009277

Hammer GL, Hansen JW, Phillips JG et al (2001) Advances in application of climate prediction in agriculture. Agric Syst 70(2–3):515–553. https://doi.org/10.1016/S0308-521X(01)00058-0

ISPA (n.d.) Precision ag definition. International Society of Precision Agriculture, Monticello IL. https://www.ispag.org/about/definition. Accessed 19 May 2022

Jarvis D (2020) The AI talent shortage isn't over yet. Deloitte Insights, 30 September. https://www2.deloitte.com/us/en/insights/industry/technology/ai-talent-challenges-shortage.html.   Accessed 20 May 2022

Jimenez A-F, Ortiz BV, Bondesan L et al (2021) Long short-term memory neural network for irrigation management: a case study from southern Alabama, USA. Precision Agric 22:475–492. https://doi.org/10.1007/s11119-020-09753-z

Kamil YA (2019) Will AI's development be hindered by a talent shortage in academia? Study International, 14 October. https://www.studyinternational.com/news/ai-professionals-talent-sho rtage/. Accessed 20 May 2022

Kant I (1965) Grundlegung zur Metaphysik der Sitten. Felix Meiner Verlag, Hamburg

Kant I (1990) Metaphysische Anfangsgründe der Tugendlehre. Felix Meiner Verlag, Hamburg

Keating GC (2018) Principles of risk imposition and the priority of avoiding harm. Revus 36:7–39. https://doi.org/10.4000/revus.4406

Kettenburg AJ, Hanspach J, Abson DJ, Fischer J (2018) From disagreements to dialogue: unpacking the Golden Rice debate. Sustain Sci 13:1469–1482. https://doi.org/10.1007/s11625-018-0577-y

Klemm T, McPherson RA (2017) The development of seasonal climate forecasting for agricultural producers. Agric for Meteorol 232:384–399. https://doi.org/10.1016/j.agrformet.2016.09.005

Leisinger K (2018) Die Kunst der verantwortungsvollen Führung. Haupt Verlag, Bern

Lemos MC, Dilling L (2007) Equity in forecasting climate: can science save the world's poor? Sci Public Policy 34(2):109–116. https://doi.org/10.3152/030234207X190964

Marcus JL, Sewell WC, Balzer LB, Krakower DS (2020) Artificial intelligence and machine learning for HIV prevention: emerging approaches to ending the epidemic. Curr HIV/AIDS Rep 17(3):171–179. https://doi.org/10.1007/s11904-020-00490-6

Maxmen A (2019) Can tracking people through phone-call data improve lives? Nature, 29 May. https://www.nature.com/articles/d41586-019-01679-5. Accessed 20 May 2022

Nagata JM, Magerenge RO, Young SL et al (2012) Social determinants, lived experiences, and consequences of household food insecurity among persons living with HIV/AIDS on the shore of Lake Victoria Kenya. AIDS Care 24(6):728–736. https://doi.org/10.1080/09540121.2011.630358

Nature (2022) Nature addresses helicopter research and ethics dumping. 2 June. https://www.nature.com/articles/d41586-022-01423-6. Accessed 30 May 2022

Owen R, Stilgoe J, Macnaghten P, Gorman M et al (2013) A framework for responsible innovation. In: Owen R, Bessant J, Heintz M (eds) Responsible innovation: managing the responsible emergence of science and innovation in society. John Wiley & Sons, Chichester, pp 27–50. https://doi.org/10.1002/9781118551424.ch2

Rosman B, Carman M (2021) Why AI needs input from Africans. Quartz Africa, 25 November. https://qz.com/africa/2094891/why-ai-needs-input-from-africans/. Accessed 20 May 2022

Scalable Cooperation (n.d.) Project Deep Empathy. School of Architecture and Planning, Massachusetts Institute of Technology. https://www.media.mit.edu/projects/deep-empathy/overview/. Accessed 18 May 2022

Schroeder D, Chatfield K, Muthuswamy V, Kumar NK (2021) Ethics dumping: how not to do research in resource-poor settings. Academics Stand Against Poverty 1(1):32–55. http://journalasap.org/index.php/asap/article/view/4. Accessed 20 May 2022

Schroeder D, Chatfield K, Singh M et al (2019) Equitable research partnerships: a global code of conduct to counter ethics dumping. Springer Nature, Cham, Switzerland. https://doi.org/10.1007/978-3-030-15745-6

Sen A (1983) Poverty and famines: an essay on entitlement and deprivation. Oxford University Press, New York

Tegli JK (2017) Seeking retrospective approval for a study in resource-constrained Liberia. In: Schroeder D, Cook J, Hirsch F et al (eds) Ethics dumping. SpringerBriefs in Research and Innovation Governance. Springer, Cham, pp 115–119. https://doi.org/10.1007/978-3-319-64731-9_14

2030Vision (2019) AI & the Sustainable Development Goals: the state of play. SustainAbility, London. https://www.sustainability.com/globalassets/sustainability.com/thinking/pdfs/2030vision-stateofplay.pdf. Accessed 20 May 2022

TRUST (2018) Global Code of Conduct for Research in Resource-Poor Settings, https://doi.org/10.48508/GCC/2018.05

UN (n.d.a) Goals: 9 Build resilient infrastructure, promote inclusive and sustainable industrialization and foster innovation. United Nations Department of Economic and Social Affairs: Sustainable Development. https://sdgs.un.org/goals/goal9. Accessed 20 May 2022

UN (n.d.b) The 17 goals. United Nations Department of Economic and Social Affairs: Sustainable Development. https://sdgs.un.org/goals. Accessed 18 May 2022

UN (2015) Transforming our world: the 2030 Agenda for Sustainable Development. Resolution adopted by the General Assembly on 25 September 2015. Res 70/1, 21 October. https://documents-dds-ny.un.org/doc/UNDOC/GEN/N15/291/89/PDF/N1529189.pdf?OpenElement. Accessed 18 May 2022

UNCTAD (2021) Inequalities threaten wider divide as digital economy data flows surge. UN Conference on Trade and Development, Geneva. https://unctad.org/news/inequalities-threaten-wider-divide-digital-economy-data-flows-surge. Accessed 20 May 2022

University of Pretoria (2018) Artificial intelligence for Africa: an opportunity for growth, development, and democratisation. Access Partnership. https://www.accesspartnership.com/cms/access-content/uploads/2018/11/WP-AI-for-Africa.pdf. Accessed 20 May 2022

UNSDG (n.d.) Leave no one behind. UN Sustainable Development Group. https://unsdg.un.org/2030-agenda/universal-values/leave-no-one-behind. Accessed 20 May 2022

Van Niekerk J, Wynberg R. (2018) Human food trial of a transgenic fruit. In: Schroeder D, Cook J, Hirsch F et al (eds) Ethics dumping. SpringerBriefs in Research and Innovation Governance. Springer, Cham, pp 91–98. https://doi.org/10.1007/978-3-319-64731-9_11

Vinuesa R, Azizpour H, Leite I et al (2020) The role of artificial intelligence in achieving the Sustainable Development Goals. Nat Commun 11:233. https://doi.org/10.1038/s41467-019-141 08-y

Weckbecker K (2018) Nicht schaden – vorsichtig sein – heilen. MMW Fortschr Med 160:36. https://doi.org/10.1007/s15006-018-0481-5

Wesolowski A, Buckee CO, Bengtsson L et al (2014) Commentary: containing the Ebola outbreak: the potential and challenge of mobile network data. PLoS Curr 6. https://doi.org/10.1371/currents.outbreaks.0177e7fcf52217b8b634376e2f3efc5e

Xinhua (2018) Greed is biggest obstacle to achieving fair societies, professor says at UN. Xinhuanet, 10 July. http://www.xinhuanet.com/english/2018-07/10/c_137313107.htm. Accessed 20 May 2022

Chapter 9
The Ethics of Artificial Intelligence: A Conclusion

Abstract The concluding chapter highlights broader lessons that can be learned from the artificial intelligence (AI) cases discussed in the book. It underlines the fact that, in many cases, it is not so much the technology itself that is the root cause of ethical concerns but the way it is applied in practice *and* its reliability. In addition, many of the cases do not differ radically from ethics cases related to other novel technologies, even though the use of AI can exacerbate existing concerns. Ethical issues can rarely be resolved to everybody's full satisfaction, not least because they often involve the balancing of competing goods. What is essential is space for human reflection and decision-making within the use of AI. Questions about what we can and should do, why we should act in particular ways and how we evaluate the ethical quality of our actions and their outcomes are part of what it means to be human. Even though Immanuel Kant believed that a good will is the only thing in the world that is ethical per se, a good will alone does not suffice where complex consequences may not be obvious. The complex nature of AI systems and their interaction with their human, social and natural environment require constant vigilance and human input.

Keywords AI ethics · Socio-technical systems · AI ecosystem · Solutions · Mitigation

This book of case studies on ethical issues in artificial intelligence (AI), and strategies and tools to overcome them, has provided an opportunity for learning about AI ethics. Importantly, it has also shown that AI ethics does not normally deal with clear-cut cases. While some cases provide examples of events that are obviously wrong from an ethical perspective, such cases are often about the *reliability* of the technology. For instance, it is obvious that AI-enabled robots should not present health, safety and security risks for users such as the death of a passenger in a self-driving car, or the smart-home system which allowed a man-in-the-middle attack. More difficult are cases where deliberation on the ethical pros and cons does not provide an immediate answer for the best approach—for instance, where robot use in elderly care reduces pressure on seriously overstretched staff but outsources important human contact

© The Author(s) 2023
B. C. Stahl et al., *Ethics of Artificial Intelligence*,
SpringerBriefs in Research and Innovation Governance,
https://doi.org/10.1007/978-3-031-17040-9_9

to machines, or where sex robots can be seen as violating the dignity of humans (especially girls and women) but at the same time helping realise sexual rights.

Looking at the cases across the different example domains in this book, one can make some general observations. The first refers to the application context of AI. Our case studies have aimed to be grounded in existing or realistic AI technologies, notably currently relevant machine learning. The ethical relevance of the cases, however, is almost always linked to the way in which the machine learning tool is applied and integrated into larger systems. The ethical concerns, then, are not focused on AI but on the way in which AI is used and the consequences this use has. For instance, the unfair dismissal case (Chap. 7) and the gender bias case (Chap. 2) are about the application of AI. Both the dismissal of staff without human input into the sequence and the training of AI devices on gender-biased CVs are about the use of AI. This is not to suggest that AI is an ethically neutral tool, but rather to high-light that the broader context of AI use, which includes existing moral preferences, social practices and formal regulation, cannot be ignored when undertaking ethical reflection and analysis.

This raises the question: how do AI ethics cases differ from other cases of technology ethics? As a first approximation it is probably fair to say that they usually do not differ radically. Many of the ethics case studies we present here are not fundamentally novel and we do not introduce issues that have never been considered before. For instance, the digital divide discussed in Chap. 8 has been debated for decades. However, the use of AI can *exacerbate* existing concerns and heighten established problems.

AI in its currently predominant form of machine learning has some characteristics that set it apart from other technologies, notably its apparent ability to classify phenomena, which allows it to make or suggest decisions, for example when an autonomous vehicle decides to brake because it classifies an object as an obstacle in the road, or when a law enforcement system classifies an offender as likely to commit a further crime despite a model rehabilitation record. This is often seen as an example of AI autonomy. It is important, however, to see that this autonomy is not an intrinsic part of the machine learning model but an element of the way it is integrated into the broader socio-technical system, which may or may not allow these classifications in the model to affect social reality. Autonomy is thus a function not of AI, but of the way in which AI is implemented and integrated into other systems. Ibrahim Diallo might not have been dismissed by a machine and escorted from the company building like a thief (see Chap. 7) if the AI system had been more transparent and required more human input into the dismissal process.

Indeed, another characteristic of current AI based on neural networks is their opacity. It is precisely the strength of AI that it can produce classifications without humans having to implement a model; that is, the system develops its own model. This machine learning model is frequently difficult or impossible for humans to scrutinise and understand. Opacity of this kind is often described as a problem and various approaches around explainable AI are meant to address it and give meaningful insight into what happens within an AI system. This raises questions about what constitutes explainability and explanations more broadly, including questions about the explainability of ethical decisions: questions that may open up new avenues in moral philosophy. And while explainability is generally agreed to be an important aspect of AI ethics, one should concede that most individuals have as little understanding of how their internal combustion engine or microwave oven works as they have of the internal workings of an AI system they are exposed to. For internal combustion engines and microwave ovens, we have found ways to deal with them that address ethical concerns, which raises the question: how can similar approaches be found and implemented for AI?

A final characteristic of current AI systems is the need for large data sets in the training and validating of models. This raises questions about ownership of and access to data that relate to the existing distribution of economic resources, as shown in Chap. 4. As data sets often consist of personal data, they may create the potential for new threats and aggravate privacy and data protection harms. This may also entrench power imbalances, giving more power to those who control such information. Access to data may also be misused to poison models, which can then be used for nefarious purposes. But while AI offers new mechanisms to misuse technology, misuse itself is certainly not a new phenomenon.

What overarching conclusions can one draw from this collection of cases of ethically problematic uses of AI and the various interpretations of these issues and proposed responses and mitigation strategies?

A first point worth highlighting is that human interaction typically results in ethical questions. Adding AI to human interaction can change the specific ethical issues, but will not resolve all ethical issues or introduce completely unexpected ones. Ethical reflection on questions of what we can and should do, why we should act in particular ways and how we evaluate the ethical quality of our actions and their outcomes are part of what it means to be human. Even though Immanuel Kant believed that a good will is the only thing in the world that is ethical per se, a good will alone does not suffice where complex consequences may not be immediately obvious. For instance, as shown in Chap. 8 about AI for Good, the most vulnerable populations might be hit harder by climate change, rather than helped, as a result of the use of AI-based systems. This was the case with small-scale farmers in Brazil and Zimbabwe who were not granted credit to cope with climate change by bank managers who had access to forecasts from seasonal climate prediction. Likewise, seasonal workers in Peru were laid off earlier based on seasonal climate forecasting. In these cases, helicopter research to aid vulnerable populations in resource-limited settings ought to be avoided, as local collaborators are likely to be in a better position to predict impacts on vulnerable populations.

Ethical issues can rarely be resolved to everybody's full satisfaction, not least because they often involve the balancing of competing goods. AI raises questions such as how to balance possible crime reduction through better prediction against possible discrimination towards disadvantaged people. How do we compare access to novel AI-driven tools with the ability and motivations of the tool holders to benefit from the use of our personal data? Or what about the possibility of improving medical diagnoses amid crippling human resource shortages versus the downsides of automated misdiagnosis? Can an uncertain chance of fighting a pandemic through AI analysis justify large-scale data collection? How could one justify the deployment of AI in resource-limited areas in the light of the intrinsic uncertainty and unpredictability of the consequences this may have on different parts of the population? What about the elderly lady whose only companion is a pet robot? All our cases can be described in terms of such competing goods, and it is rarely that a simple response can be given. The conclusion to be drawn from this is that awareness of ethical issues and the ability and willingness to reflect on them continuously need to be recognised as a necessary skill of living in technology-driven societies.

Another conclusion to be drawn from our examples is that the nature of AI as a system needs to be appreciated and included explicitly in ethical reasoning. AI is never a stand-alone artefact but is always integrated into larger technical systems that form components of broader socio-technical systems ranging from small-scale local systems in individual organisations all the way up to global systems such as air traffic control or supply chains. This systemic nature of AI means that it is typically impossible to predict the consequences of AI use accurately. That is a problem for ethical theory, which tends to work on the assumption that consequences of actions are either determined or at least statistically distributed in a way that can be accurately described. One consequence of this lack of clear causal chains in large-scale socio-technical systems is that philosophy could aim to find new ways of ethical reflection of systems.

In practice, however, as our description of the responses to the cases has shown, there is already a significant number of responses that promise to be able to lead to a better understanding of AI ethics and to address ethical issues. These range from individual awareness, AI impact assessments, ethics-by-design approaches, the involvement of local collaborators in resource-limited settings and technical solutions such as those linked to AI explainability, all the way to legal remedies, liability rules and the setting up of new regulators. None of these is a panacea which can address the entire scope of AI ethics by itself, but collectively and taken together they offer a good chance to pre-empt the significant ethical problems or prevent them from having disastrous consequences. AI ethics as systems ethics provides a set of ethical responses. A key challenge that we face now is to orchestrate existing ethical approaches in a useful manner for societal benefit.

AI ethics as an ethics that takes systems theory seriously will need to find ways to bring together the approaches and responses to ethical challenges that we have presented. The responses and mitigation strategies put forward here do not claim to be comprehensive. There are many others, including professional bodies, standardisation, certification and the use of AI incident databases, to name but a few. Many of these already exist, and some are being developed and tailored for their application to AI. *The significant challenge will be to orchestrate them in a way that is open, transparent and subject to debate and questioning, while at the same time oriented towards action and practical outcomes.* Regulation and legislation will likely play a key role here, for example the European Union's Artificial Intelligence Act proposal, but other regulatory interventions, such as the creation of AI regulators, may prove important (Stahl et al. 2022). However, it is not just the national and international policymakers that have to play a role here. Organisations, industry associations, professional bodies, trade unions, universities, ethics committees, the media and civil society need to contribute. All these activities are based on the effort and contributions of individuals who are willing to participate in these efforts and prepared to reflect critically on their actions.

Tackling AI ethics challenges is no simple matter, and we should not expect to be able to solve all ethical issues. Instead, we should recognise that dealing with ethics is part of what humans do and that the use of technology can add complexity to traditional or well-known ethical questions. We should furthermore recognise that AI ethics often cannot be distinguished from the ethics of technology in general, or from ethical issues related to other digital and non-digital technologies. But at the same time, it has its peculiarities that need to be duly considered.

Our aim in this book has been to encourage reflection on some interesting cases involving AI ethics. We hope that the reader has gained insights into dealing with these issues, and understands that ethical issues of technology must be reflected upon and pursued with vigilance, as long as humans use technology.

Reference

Stahl BC, Rodrigues R, Santiago N, Macnish K (2022) A European agency for artificial intelligence: protecting fundamental rights and ethical values. Comput Law Secur Rev 45:105661. https://doi.org/10.1016/j.clsr.2022.105661

Index